American Institute of Chemical Engineers

Annual Meeting of the American Electrophoresis Society

Topical Conference at the
2007 AIChE Annual Meeting

November 4-9, 2007
Salt Lake City, Utah, USA

Printed from e-media with permission by:

Curran Associates, Inc.
57 Morehouse Lane
Red Hook, NY 12571
www.proceedings.com

ISBN: 978-1-60423-827-3

Some format issues inherent in the e-media version may also appear in this print version.

ISBN: 978-1-60423-827-3

American Institute of Chemical Engineers
Proceedings
Three Park Avenue
New York, NY 10016-5991
Phone: 212-591-8100

www.aiche.org

American Institute of Chemical Engineers

Annual Meeting of the American Electrophoresis Society
2007

TABLE OF CONTENTS

11a Multiplexed Analysis Of Microbial Proteomes

Carol S. Giometti, Biosciences Division, Argonne National Laboratory, 9700 South Cass Avenue, Building 202, Room B117, Argonne, IL 60439

A broad range of protein separation and analytical methods are now available for in depth characterization of the protein complement of microbial cells. A multiplexed approach utilizing a variety of electrophoresis, chromatography, and mass spectrometry methods is being used to study the response of bacterial to different growth conditions at the level of total steady state protein content, membrane protein composition, protein-protein interactions, and post-translation modification. As an example of the comprehensive data sets produced using this multiplexed analysis strategy, proteome studies of *Shewanella oneidensis*, an exquisitely responsive metal-reducing bacterium, are revealing changes in the steady-state abundance of specific proteins as well as post-translational modifications when cells are grown with different terminal electron acceptors. Comparison of *Shewanella oneidensis* total and membrane protein fractions after separation by two-dimensional gel electrophoresis (2DE) is used to detect changes in steady-state protein levels, and protein identifications are determined by tandem mass spectrometry. Parallel analyses using two-dimensional liquid chromatography coupled to tandem mass spectrometry (2DLC/MS-MS) provides a broader inventory of protein constituents together with relative abundance. The data produced by such multiplexed analysis necessitates the use of a relational database system for data analysis, management, and integration. To date, over 500 *Shewanella oneidensis* proteins have been identified using 2DE together with tandem mass spectrometry with an additional 500 identified by 2DLC/MS-MS. Serine/threonine phosphorylation of a small subset of these proteins has been identified and preliminary studies suggest that differential phosphorylation of specific *Shewanella oneidensis* proteins correlates with electron acceptor availability.

11b Quantitative Phosphoproteomics and in-Depth Protein Profiling of Melanoma Cell Lines

Katheryn Resing, Department of Chemistry and Biochemistry, University of Colorado at Boulder, 3990 Carlock Drive, Boulder, CO 80309

Analysis of kinase signaling processes in diseases such as melanoma requires quantitative information on the phosphorylation sites of a large number of proteins. In addition, the proteins themselves must be quantified, in order to distinguish changes in stoichiometry vs changes in protein levels (for instance, from relocalization between cell compartments). Methods involving enrichment of phosphopeptides are difficult to reproduce and do not capture all phosphopeptides. Instead, we have developed a shotgun proteomics protocol exploiting scanning for phosphate loss during negative ion MS/MS in an ABI 4000QTrap for phosphopeptides (-79 scanning). This is combined with parallel profiling of the proteins on a ThermoFisher LTQ/Orbitrap. We are using this approach to analyze MAPK kinase signaling in melanoma cell lines; this pathway is constitutively activated in ~75% of melanomas. Our current studies compare protein profiles and phosphopeptide features in a metastatic cell line treated with MAPK kinase inhibitor vs the control cells, extracting the soluble proteins and resolving them on monoQ into ~50 fractions. New computational methods based on OpenMS were developed to match features between individual LC/MS analyses of each fraction, allowing quantification of the phosphopeptide ion intensities in the negative ion QTrap scans. The number of phosphopeptide features detected in the QTrap (~5000) was similar to the number of proteins identified in the Orbitrap profiles (~4,300 proteins, with evidence of saturation sampling). These numbers suggest that we are likely sampling almost all the detectable phosphorylation sites, because they are consistent with estimates that ~1/3 of proteins are phosphorylated and that the average phosphoprotein has about 3 phophorylation sites. Of the QTrap detected phosphopeptide features, ~6% changed intensity upon treatment of the cells with MAPK kinase inhibitor. MS/MS and MS3 data collected in the same experiments provides sufficient information to identify some of these peptides, and many of these are consistent with known MAPK kinase pathway targets. We are now developing methods to target the remaining unidentified phosphopeptide features.

Signal-to-Noise and Quantitative Proteomics Using Difference Gel Electrophoresis.

David B. Friedman

The complex methodology used in many large-scale quantitative proteomics experiments often dictates an experimental design with small sample size and limited repetition. Variation in a complex dataset may hopefully arise from changes caused by the experimental condition(s), but could also arise due to technical noise (poor sample prep, run-to-run variation) and biological noise (normal differences between samples, especially present in clinical samples). Difference Gel Electrophoresis (DIGE) is particularly well-suited for quantitative proteomics studies. DIGE uses a common pooled-sample internal standard to coordinate multiple 2D gels, each of which also contains two multiplexed samples from a large sample set representing independent (biological) replicates from multiple conditions. Multivariate statistical analyses such as Principle Component Analysis (PCA) and unsupervised Hierarchical Clustering (HC) can now be applied to complex DIGE datasets. These tools provide a global perspective on multivariable DIGE datasets that can assess whether the variation in the system describes the biological signal, rather than being derived from technical/biological noise whereby 'significant' changes may arise stochastically. Examples will be shown from experiments containing samples from micro-organisms, tissue culture and clinical samples, where these tools were instrumental in demonstrating sample outliers, fouled samples, as well as variation in sample preparation that overrides the variation from biological treatment (despite standard biological tests for sample validity).

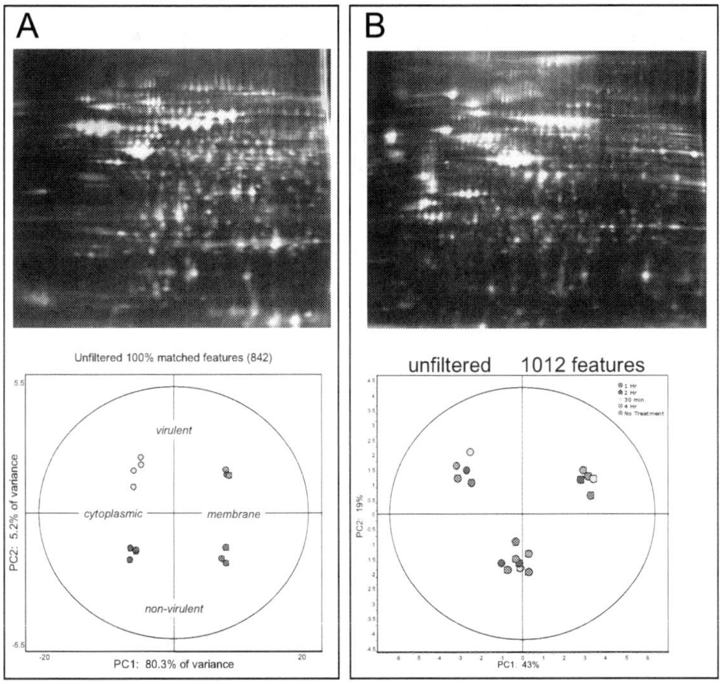

Panel A depicts a representative DIGE gel from an 8-gel experiment resolving 16 samples from four conditions (N=4 for each). The Principle Component Analysis (PCA) shown below suggests that the major source of variation (PC1) arises from expected differences between cytoplasmic and membrane fractions, with a smaller PC2 component describing variation between a virulent and non-virulent strain of H. pylori. **Panel B** shows a similar scenario for human fibroblasts treated with drug over 5 timepoints (N=4 for each timepoint; 10 DIGE gels total). In this case, PCA clearly demonstrates that the major source of variation is arising between the replicate sample preparations, and not due to drug treatment.

68a Microfluidic Delivery of Small Molecules into Mammalian Cells Based on Hydrodynamic Focusing

Fen Wang[1], Hao Wang[2], Jun Wang[3], Hsiang-Yu Wang[4], Peter Rummel[3], Suresh Garimella[2], and **Chang Lu**[5]. (1) Agricultural and Biological Engineering, Purdue Univerisity, West Lafayette, IN 47907, (2) Mechanical Engineering, Purdue University, West Lafayette, IN 47907, (3) Agricultural and Biological Engineering, Purdue University, West Lafayette, IN 47907, (4) Chemical Engineering, Purdue University, West Lafayette, IN 47906, (5) Agriculture and Biological Engineering, Chemical Engineering, Purdue University, West Lafayette, IN 47906

Cell-based assays on a microfluidic platform have received increasing interest from chemists, biologists, and engineers. Microfluidics-based cell assays offer high levels of automation and integration, and allow multiple assays to be run in parallel, based on reduced sample volumes. These characteristics make them attractive for studies associated with drug discovery. Controlled delivery of drug molecules or other exogenous materials into cells is a critical issue that needs to be addressed before microfluidics can serve as a viable platform for drug screening and studies. In this study, we report the application of hydrodynamic focusing for controlled delivery of small molecules into cells immobilized on the substrate of a microfluidic device. We delivered calcein AM which was permeant to the cell membrane into cells, and monitored its enzymatic conversion into fluorescent calcein during and after the delivery. Different ratios of the sample flow to the side flow were tested to determine how the conditions of hydrodynamic focusing affected the delivery. A 3D numerical model was developed to help understand the fluid flow, molecular diffusion and delivery based on hydrodynamic focusing. The results from the simulation indicated that the amount of calcein AM delivered was strongly correlated with its concentration distribution on the cell surface, which was in turn determined by the hydrodynamic focusing. We have demonstrated the feasibility of using microfluidic hydrodynamic focusing for potentially quantitative delivery of exogenous molecules into cells at the single cell or subcellular level. We expect that our technique will pave the way to high-throughput drug screening and delivery on a microfluidic platform.

68b Development Of A Microfluidic Genetic Bead Hybridization Platform For The Rapid Identification Of DNA

Jason E. Gordon, Satyajyoti Senapati, Diana Hou, Shramik Sengupta, and Hsueh-Chia Chang. Chemical & Biomolecular Engineering, University of Notre Dame, 182 Fitzpatrick Hall, Notre Dame, IN 46556

Ballast water is a necessary component of the shipping industry where it is employed for weight balancing. Unfortunately, such water has been implicated as one of the chief means of aquatic invasive species introduction worldwide. Consequently, the rapid genetic identification of organisms within a ballast water sample is of interest. Traditionally, genetic identification is accomplished using conventional microarray technology. Such technology is attractive in its ability to offer large oligomer library scans. However, the existence of large diffusion distances between target DNA and probe DNA results in lengthy assay times, making these microarrays impractical for rapid identification purposes.

Recently, significant attention has been given to developing a microfluidic chip that can be utilized for conducting DNA-DNA hybridization studies. We report here on the development of a novel microfluidic chip that employs a genetic bead hybridization platform. This chip has been fabricated such that oligomer functionalized beads are packed within a miniature hybridization chamber. This packing is accomplished through the fabrication of porous filters embedded within the microchannels that are connected to the hybridization chamber. Heat-denatured DNA is then passed through the microchannels and into the genetic bead packed bed region. The implementation of this genetic bead packed bed region is advantageous in that the target DNA is brought into close contact with the oligomers on the surface of the beads. Additionally, the surface area available for hybridization is greatly increased. Finally, the detection of positive hybridization is accomplished through fluorescence.

Extended Abstract:

Microfluidic Design and Flow Control in Microbioreactor Array with a Common Feed Inlet

Yuan Wen, Shang-Tian Yang
The Department of Chemical and Biomolecular Engineering
The Ohio State University

Polymeric microfluidic channels have high potentials in the use of disposable diagnostic and analytical devices due to the advantages of low cost of materials, small reagent consumption and highly parallel experimentation capability. However, at micrometer scale, capillary effect plays a prominent role in liquid flow at the startup phase, especially creating problems in filling multiple hydrophobic channels with abruptly changing boundaries from one common inlet. This problem may even occur in just two parallel poly(dimethylsiloxane) (PDMS) microchannels fabricated with multi-layer soft lithography. This study presents a preliminary investigation for this phenomenon. By theoretical studies based on microfluidic models for pressure drop, uniform flow may be achieved by specifically designed geometries for microchannels and/or by surface modification. A novel and convenient solution was achieved by using fibrous matrices to fill microchannels, thus increasing flow resistance in order to balance unwanted capillary pressures for achieving uniform flow. The result from this research demonstrated the importance of material selection, design and process for fabricating microfluidic devices, which may potentially benefit high-throughput experimentations using plastic microchips.

RESULTS

1 Non-uniform flows in 10×1 array device

The flow in the 10×1 array device was recorded in a video, and images at different time points were captured from the video to demonstrate how the liquid distributed throughout the 10 channels (Figure 1). A non-uniform flow distribution was observed. The flow rate gradually decreased from the middle channels to the channels at the two sides. This is one level of the non-uniform flow in such a device. Another even more severe non-uniform flow was observed for a video recorded with a longer time (Figure 2). In this scenario, the non-uniform flow rate distribution was also observed. Furthermore, once one of the channels was filled with the liquid, the other 9 channels were left in idleness from then on, which is a fatal failure for the operation of the device. The intuitive design of the 10×1 array device proved naive in controlling multiple parallel flows in microfluidic devices. However, at least we could learn that each channel should have the same travel length to ensure uniform flow rates throughout all the different channels.

2 Non-uniform flow in 2×4 array device

The flow in the 2×4 array device which did not contain PET fibrous matrices in the cell culture wells was recorded in video (Figure 3). The two parallel microchannels had a symmetric design with the same travel length, but the flows were still uneven. This might be caused by imperfection of manufacture and perturbations from environment. However, the position which actually caused the flow to pause has some implications further than just imperfection of manufacture or external perturbation. As we can see in Figure 3 B and C, the pause of the flow in Line 2 occurred at the entry to a cell culture well, and this flow did not resume until the flow in Line 1 reached the entrance to another cell culture well. Furthermore, once the flow got passed in the microchannels, it seemed that the resistance became minimal. Therefore, we may deduce that the resistance in a microchannel at the entrance to a cell culture well may be higher than that at the other positions throughout the channel in the flow startup phase, for the capillary pressure at the air/liquid interface plays a dominant role in microchannels. This is confirmed by theoretical analysis. By using equation 2.8, we obtained the capillary pressure at the inlet of the first cell culture well Pc as large as 216.1 Pa, and 186.5 Pa at the entrances for the following cell culture wells, which all have a direction from air to liquid due to the hydrophobicity of the PDMS surfaces. Furthermore, after the microchannel is filled with liquid, pressure drops through the connecting channel between the cell culture well and the relay well are estimated to be just around 6.0 Pa based on equations 2.2, 2.4, and 2.5. If we assume that the liquid passes a cylinder-shaped channel in either the cell culture well or relay well, the pressure drops caused by the geometries of the wells may also be obtained with equation 4.1 as 0.003 Pa in cell culture well and 0.02 Pa in relay well, which are simply negligible compared to other pressure drops. As we may see, the capillary pressure is so dominant in the startup phase that it is even larger than the summation of all the flow resistance generated by a whole line containing 8 connecting channels, 4 cell culture wells and 3 relay wells with liquid fully filled (186.5 Pa or 216.1 Pa > 48.072 Pa). In addition, if we take inertia into consideration, we may find that extra pressure is even required to get the halted flow started compared with the line that is already with a running flow, even when both the lines reach an entrance to a cell culture well simultaneously. Therefore, it is not surprising to see how the non-uniform flow was generated in the 2×4 array microlfuidic device.

The device above did not contain the PET fibrous matrices required for the tissue culture, so further experiment was conducted to record the flow in the 2×4 array device with the presence of PET fibers filling each cell culture well (Figure 4). As we may see again, the flow was not uniform. In addition, the locations where the flows were put to halt were all at the entrances to a cell culture well. The same theoretical analysis may apply as above. Besides the pressure drops in the connecting channels and from capillarity, we may obtain the pressure drop as 12.4 Pa through the cell culture well filled with PET matrices, estimated using equations 4.2, 4.3 and 4.4. This time, again, the capillary pressure is still much larger than all the flow resistance in one fully filled line containing 8 connecting channels, 4 PET-fibrous-matrix filled cell culture wells and 3 relay wells (186.5 Pa or 216.1 Pa > 97.672 Pa).

Furthermore, it is also a concern that the manufacture errors were so bad that they caused all the non-uniform flows. In addition, the real situation for perfusion cell culture is that the flow will only start when all the cell culture wells are filled with seeded, wet PET fibrous matrices, which may even further complicate the flow resistance. Therefore, the following experiment was conducted to reveal more about the nature of the non-uniformity of the flow. A red-dye pre-wetted microfluidic 2×4 array device as shown in Figure 5 A was perfused with a green liquid. The flow sequence was designed in a way that the green liquid was first filling the device as in Figure 5B, followed by air purging (Figure 5C), and then green liquid filling again (Figure 5D). It was found that the flow might pick either Line 1 or Line 2, and the line that was at halt only occurred at the entrance to a cell culture well, demonstrating that the manufacture errors should not be the major cause for non-uniform flows, but capillary pressure plays a key role in the startup of liquid flows.

3 Uniform flow in 2×4 array device

The previous experiments indicated that non-uniform flow was mainly caused by very large capillary pressure at the entrance of cell culture well. Therefore, if we keep the design unchanged, increasing flow resistance for each microfluidic line to a value that is comparable with the capillary pressure becomes a natural resort to making the device less susceptible to failed startup. By theoretical analysis, we obtained that the pressure drop in the relay well filled with PET fibrous matrices could be as high as 27.9 Pa. Therefore, for a fully filled line containing 8 connecting channels, 4 PET-fiber filled cell culture wells, and 3 PET-fiber filled relay wells, the hydrodynamic pressure drop may add up to 181.372 Pa, which is comparable with 186.5 Pa or 216.1 Pa for a single entrance to a cell culture well. In this way, even if one line is filled faster than the other at the beginning due to small manufacture errors or external perturbation, at the entrance to a cell culture well, the flow would still "wait" for the other line to catch up due to the high hydrodynamic pressure drops along the line with fibrous matrices filled wells. With this theoretical analysis, another experiment was conducted with all the cell culture wells and relay wells filled with PET fibrous matrices, and the flow was recorded at different time points (Figure 6). In this experiment, parallel uniform flow was obtained just as the theoretical analysis implied. In order to test how stable the parallel flow was, a liquid flow using green dye was employed after using the red dye. Figure 7 demonstrates that the uniform flow was maintained in the microfluidic 2×4 array device, indicating that capillary pressure is no longer present to complicate the liquid flow when the device is fully perfused.

CONCLUSIONS

Microfluidic systems provide the enabling technologies for minimizing reagent consumption and conducting high-throughput experimentations. However, at micrometer scale, fluid behaviors may be affected by different factors than at macroscale. Capillary pressure plays a key role in microfluidic channels made of hydrophobic materials with a value that is one to two orders of magnitude larger than hydrodynamic pressure drops in simple microchannels at the startup phase. After filling, however, the capillarity no longer affects the system. This especially causes non-uniform flows for microfluidic devices with multiple hydrophobic channels. This problem may be cured by material selection, design and process optimization. In this report, based on experimental data and theoretical studies, PET fibrous matrices were employed to engineer the fluid resistance of microfluidic devices which would otherwise have non-uniform flows due to the presence of capillary pressure. The findings in this study may help researchers in the microfluidic field to better understand the major forces at this scale, so that it would benefit the design and fabrication of high-throughput microfluidic devices with hydrophobic materials.

A. t=2s
B. t=6s
C. t=8s
D. t=10s

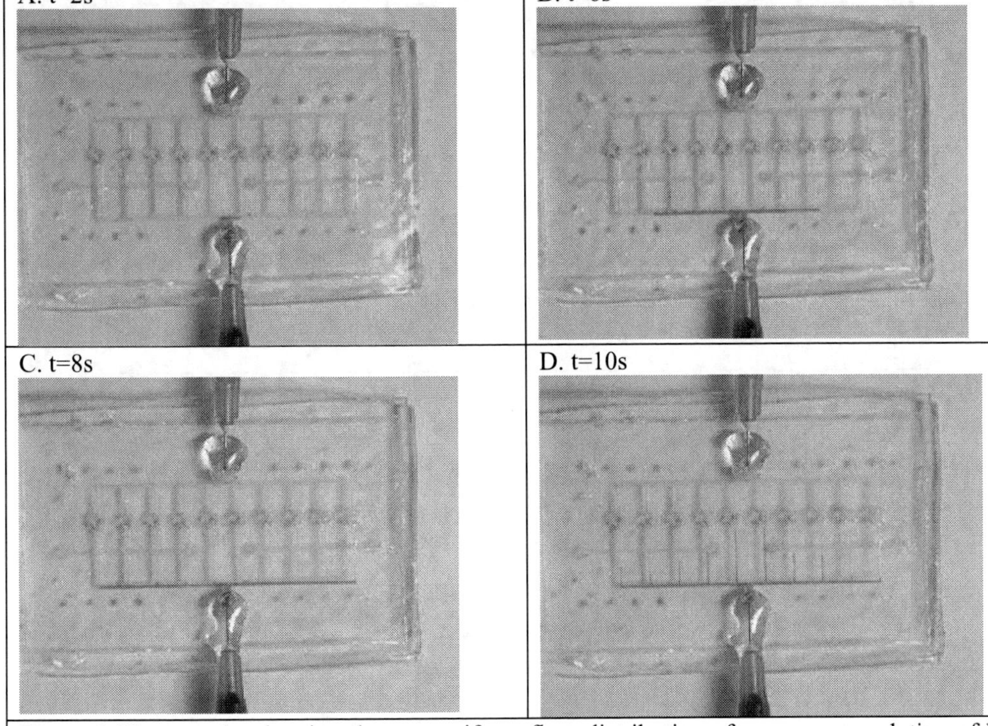

Figure 1 Photographs showing the non-uniform flow distribution of an aqueous solution of blue dye throughout 10 parallel channels in a device. The flow rate decreases from the middle channels to the channels at the two sides due to non-equal travel lengths.

A. t=1s
B. t=7s

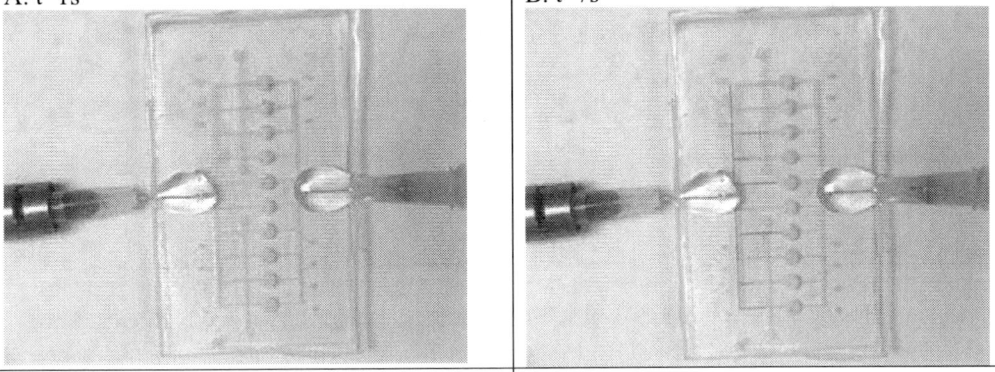

C. t=11s
D. t=20s

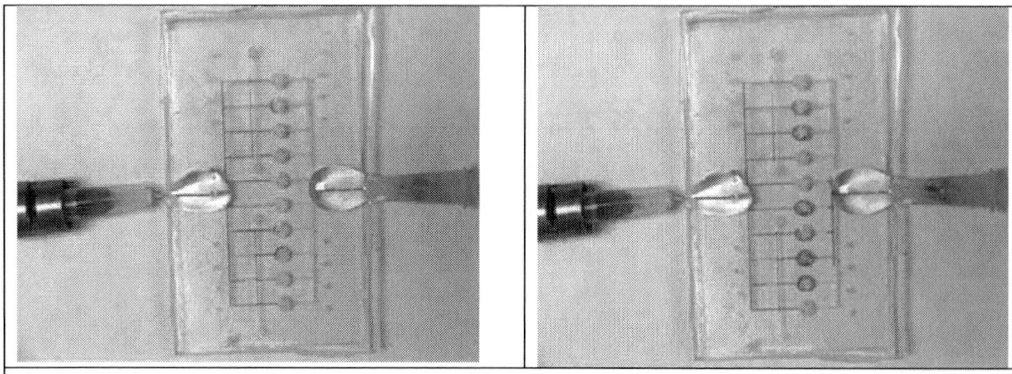

Figure 2 Photographs showing the non-uniform flow distribution in a 10×1 array device. Not only the flow rate is different throughout different channels as shown in B, but also once one of the channels is fully filled with the liquid flow, the other channels are left unable to be filled, as shown in D, which causes the idleness of the other 9 channels, determining the failure of the operation.

A. t=1s

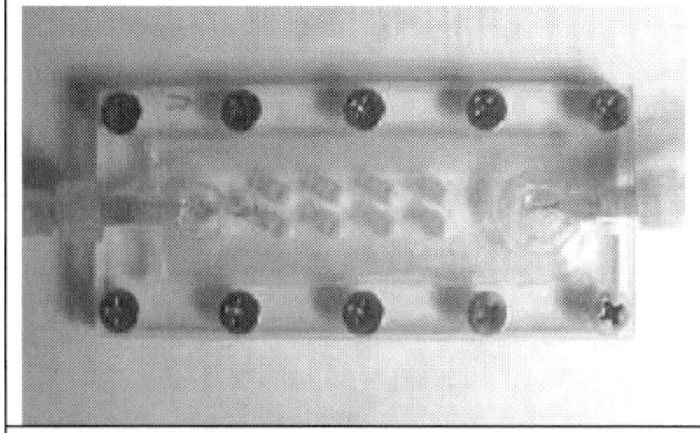

B. t=5s

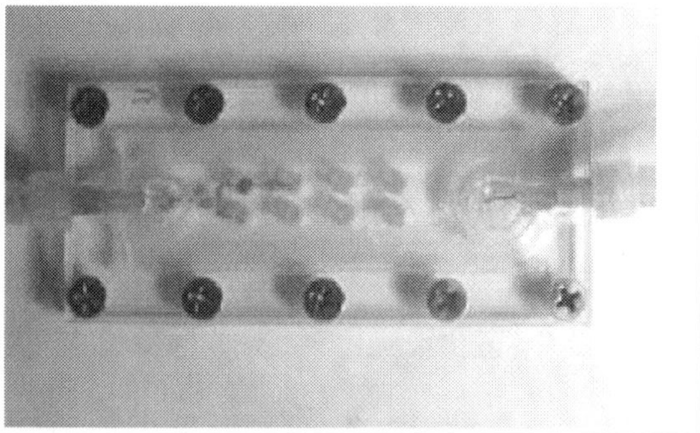

C. t=7s

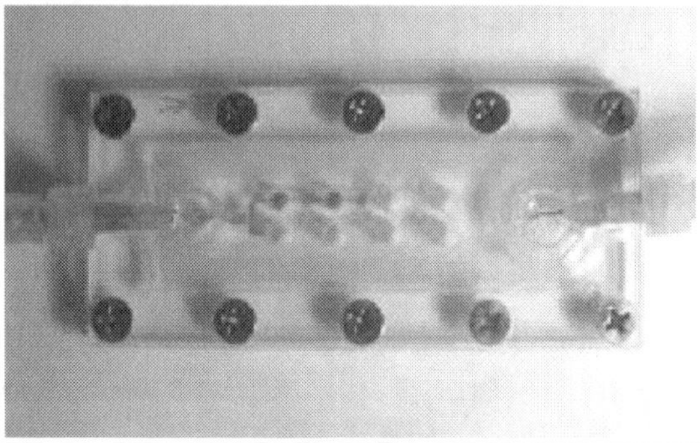

D. t=15s

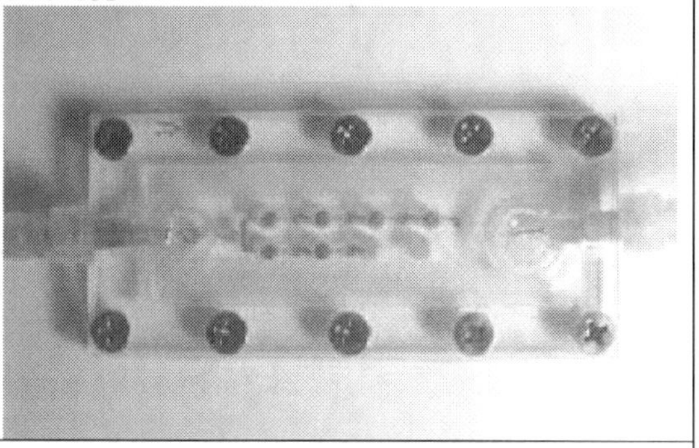

Figure 3 Photographs showing liquid flow at different time points in the 2×4 array device which has no PET fibrous matrices filled in the cell culture wells. Even with the design of two parallel channels having the same travel length for the flows, the imperfection of the manufacture and even small perturbations may make the two flows uneven. The spot which stops the flow for Line 2 is at the entrance to a cell culture well, as shown in B and C. Line 2 doesn't resume flow until the flow in the Line 1 reaches another entrance to a cell culture chamber, as shown in C. And after Line 1 is fully filled with the liquid flow, the other microchannel is "permanently" inactive at small flow rates, as seen from D.

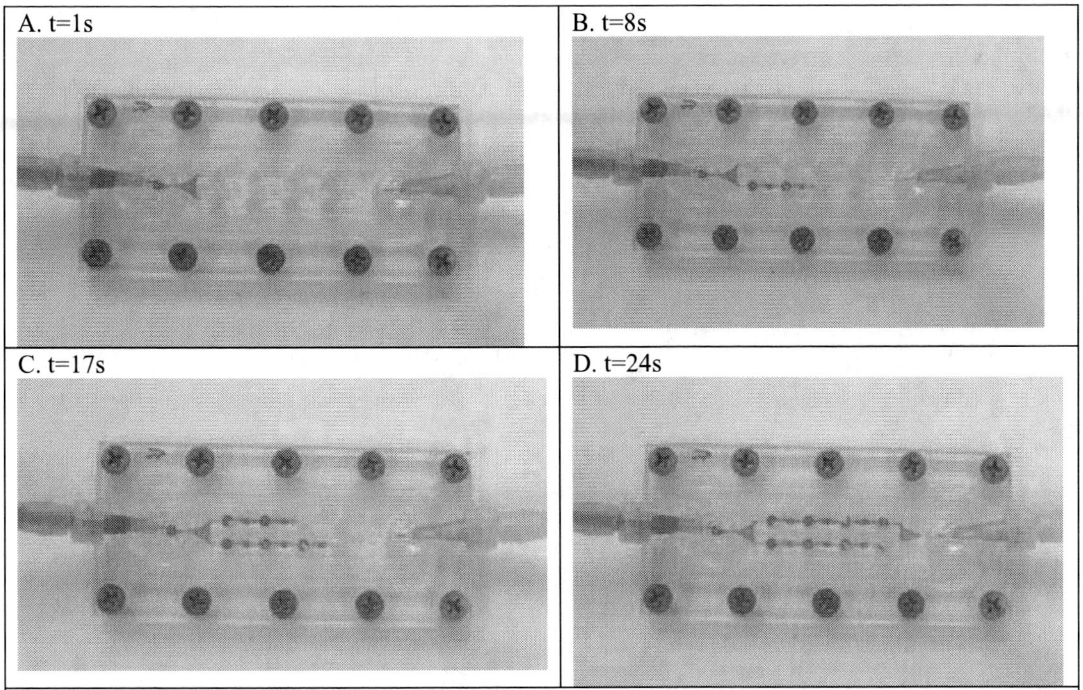

Figure 4 Photographs showing liquid flow through a 2×4 array device with cell culture wells filled with PET fibrous matrices. Non-uniform flow was also generated. The flow in Line 1 is put to halt at the entrance to the first cell culture well, where capillary pressure is dominant, as shown in B. The flow in Line 1 resumes when the flow in Line 2 reaches the entrance to a cell culture well, as shown in B and C. However, the flow in Line 2 is put to halt at the entrance to its 4th cell culture well, while the flow in Line 1 fully fills the line, leaving Line 2 "permanently" inactive, as shown in D.

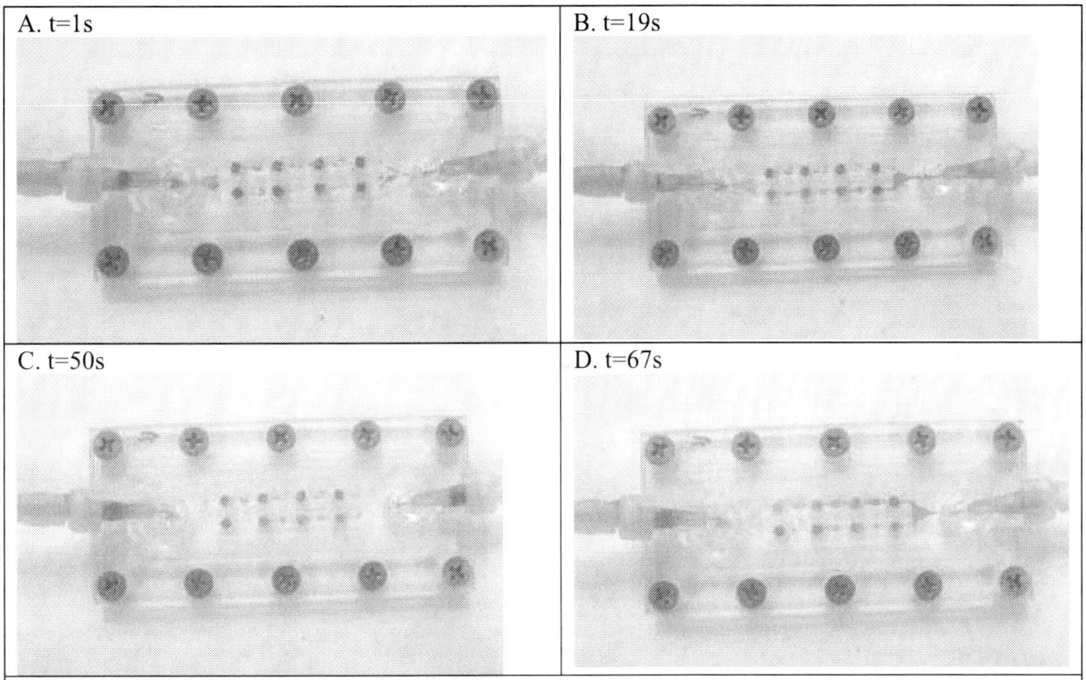

Figure 5 Photographs showing a flow indicated by a green dye passes through a red-dye pre-wetted (as in A) 2×4 array device with the cell culture wells filled with PET fibrous matrices. The green liquid is originally separated into two parts with some air sandwiched in the middle, generating a flow sequence as first filling with green liquid as in B, then air purging as in C, and finally filling with green liquid again as in D. The liquid flow either picks Line 2 or Line 1 and the halted line is at the

entrance to the first cell culture well in both cases, indicating that the manufacture errors are not the major cause for non-uniform flow, but capillarity plays a dominant role in the flow startup in the microfluidic device.

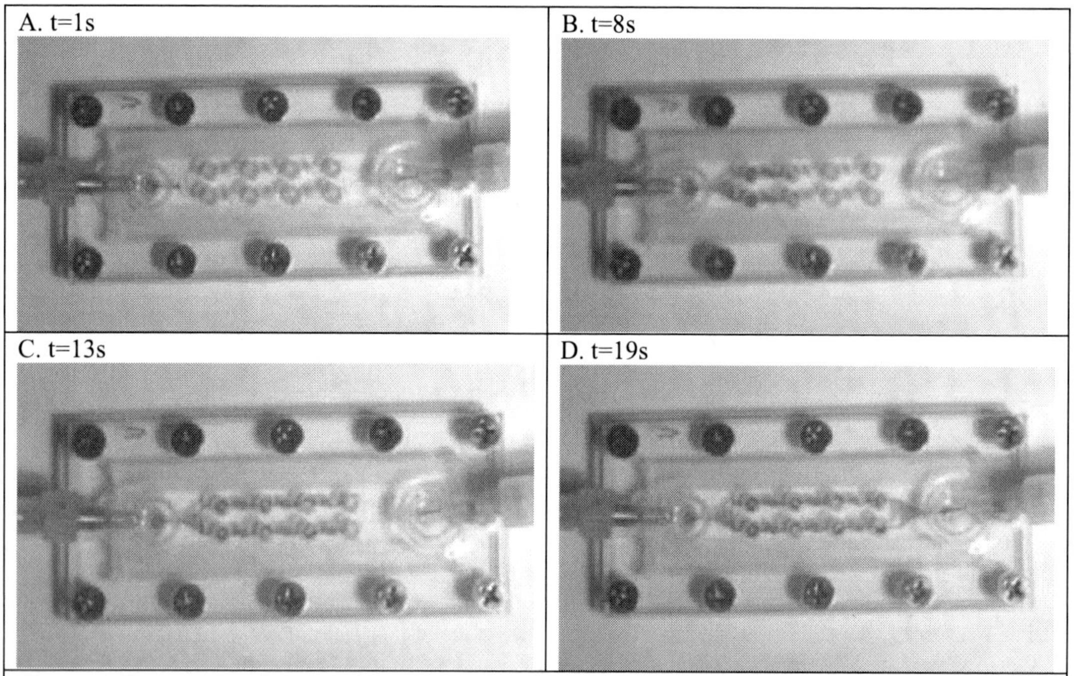

Figure 6 Photographs showing the uniform liquid flow through a 2×4 array device with both cell culture wells and relay wells filled with PET fibrous matrices. The parallel flows are achieved by increasing the hydrodynamic flow resistance for each line to a value that is comparable with capillary pressure.

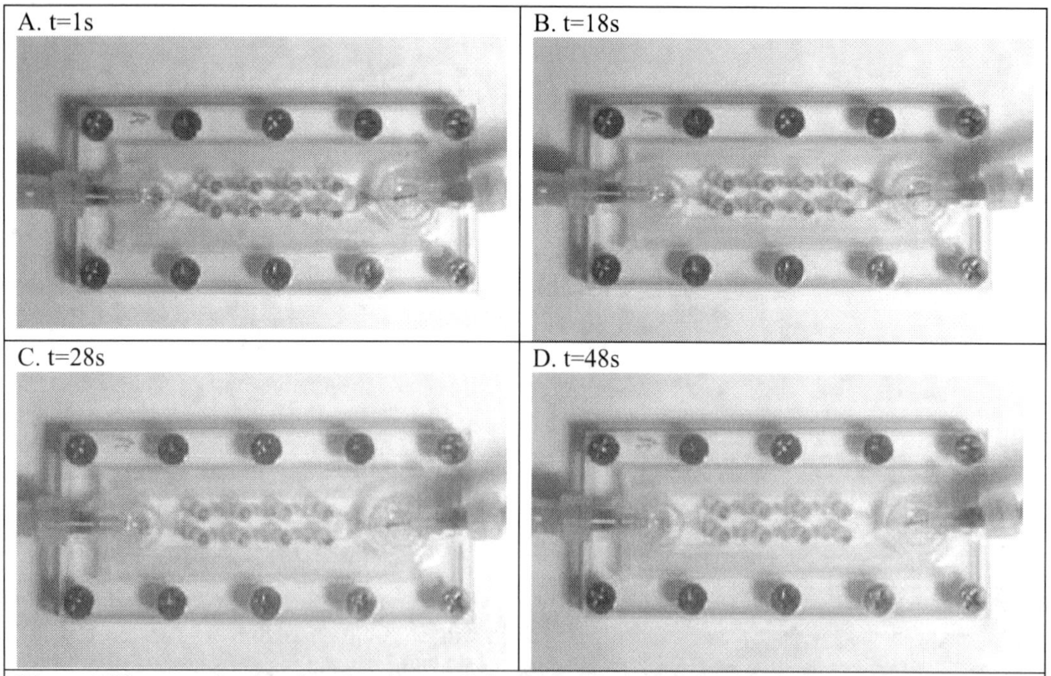

Figure 7 Photographs showing that the parallel uniform flow is maintained after changing from red liquid to the green when the 2×4 array device is fully perfused.

68d Barcoded Microparticles For Multiplexed Biomolecule Detection

Daniel C. Pregibon, Chemical Engineering, MIT, 66-053, 77 Massachusetts Ave., Cambridge, MA 02139, Mehmet Toner, Massachusetts General Hospital/Shriners Burn Hospital/Harvard Medical School, 51 Blossom Street, Boston, MA 02114, and Patrick S. Doyle, Department of Chemical Engineering, MIT, 66-053, 77 Massachusetts Ave., Cambridge, MA 02139.

High-throughput screening for genetic analysis, combinatorial chemistry, and clinical diagnostics benefits from multiplexing, which allows for the simultaneous assay of several analytes but necessitates an encoding scheme for molecular identification. Current approaches for multiplexed analysis involve complicated or expensive processes for encoding, functionalizing, or decoding active substrates (particles or surfaces) and often yield a very limited number of analyte-specific codes. We present a method based on continuous-flow lithography (Dendukuri et al. Nat. Mat. 2006) that combines particle synthesis, encoding, and probe incorporation into a single process to generate multifunctional particles bearing over a million unique codes. By using such particles, we demonstrate a multiplexed, single-fluorescence detection of DNA oligomers with encoded particle libraries that can be scanned rapidly in a flow-through microfluidic channel (Pregibon et al. Science 2007). Furthermore, we demonstrate with high specificity the same multiplexed detection using single, multiprobe particles.

Web Page: web.mit.edu/doylegroup

68e Spatial Quantification Of Abo Blood Antigen Contributions To Field And Frequency Modulated Dielectrophoresis

Prashant Reuben Daggolu and **Adrienne R. Minerick**. Dave C. Swalm School of Chemical Engineering, Mississippi State University, Swalm 258, Mississippi State University, Mississippi State, MS 39762

Dielectrophoretic manipulations are useful for applications in biological research such as separation of malarial cells and cancer cells from healthy cells. The response of red blood cells to a dielectrophoretic field is of a particular interest due to blood's easy accessibility and prominent role in medical diagnostics. Dielectrophoresis of red blood cells is being studied to assess its usefulness as a tool in portable, point-of-care, medical diagnostic microdevices. Previous research established that red blood cells translate across electric field lines to form chains in two dimensional dielectrophoretic fields. Our current research has discovered that the movement of blood cells in dielectrophoretic fields is characteristic of their ABO blood type. Utilizing edge recognition image analysis software, cell positions are tracked vertically along approximate field lines as well as horizontally across field gradients. In the present work, we investigate the dependency of dielectrophoretic field strength and frequency on the movement of positive ABO red blood cell types. Experiments were performed at five different field strengths and six different frequencies to assess and quantify cell movements and behaviors in the dielectrophoretic field. Profiles for the positive blood types have been compiled and suggest the presence or absence of ABO blood antigens have subtle impacts on cell movement in dielectrophoretic fields. This study will potentially elucidate blood type dependencies and determine optimal field strength and frequency for each blood system. These results will be important in the future development of dielectrophoretic medical diagnostic kits.

Web Page: www.che.msstate.edu/research/MDERL/

68f Microfluidic Separation Of Cardiac Cell Subpopulations

Shashi K. Murthy, Department of Chemical Engineering, Northeastern University, 360 Huntington Ave., 342 SN, Boston, MA 02115, Milica Radisic, IBBME/Chemical Engineering and Applied Chemistry, University of Toronto, 164 College Street, Room 407, Toronto, ON M5S 3G9, Canada, and Brian Plouffe, Chemical Engineering, Northeastern University, 360 Huntington Ave., 342 SN, Boston, MA 02115.

Microfluidic approaches to the separation of cardiac cell subpopulations offer significant advantages over conventional techniques such as pre-plating, cell straining, or fluorescence-activated cell sorting. These approaches are capable of handling small numbers of cells and high-resolution separation. The major application of microfluidic separation of cardiac cells is in tissue engineering, where enriched subpopulations of cells can be cultured in vitro to create functional units of cardiac tissue. This presentation will describe size- and adhesion-based approaches for microfluidic separation of cardiac cells.

68g Polymeric Microfluidic Chips In Bioanalytical Applications

Malgorzata A. Witek, Mateusz L. Hupert, and Steven A. Soper. Department of Chemistry and Center for BioModular Multi-Scale Systems, Louisiana State University, 232 Choppin Hall, Baton Rouge, LA 70803

Novel technologies are leading the way toward inexpensive diagnostic devices to perform an array of specific diagnostic tests. To meet the demands of clinical testing, devices should possess attributes such as high throughput capabilities, flexibility in the choice of assay format; multiplexing capabilities, and low cost and disposability. Microfluidic systems fabricated in polymers meet a lot of these criteria and thus are attractive materials for molecular analysis. Application of polymeric microfluidic devices developed in our group, as potential diagnostic tools will be presented. One such application involves a purification and pre-concentration of DNA from biological samples, which is a critical step in many bioassays. A device for a solid-phase nucleic acids purification (i.e., gDNA, mRNA, sequencing products) will be described as an example of high parallel processing method of sample preparation, and as front module in the integrated microfluidic systems. An integrated microfluidic chip hosting continuous flow polymerase chain reaction with ligase detection reaction (PCR/LDR), designed for colorectal cancer mutation detection will be presented. Also, the operation of the reversed transcription-polymerase chain reaction (RT-PCR) chip applicable in viral infections detection will be discussed. The specific advantages of these devices include very low cost of fabrication achievable through use of polymer as the substrate material. Low cost of the device allows for its single time, disposable use which is mandatory in the processing of medical and forensic samples.

Miniaturization of Warfarin Metabolism Genotyping by Melting Analysis using Asymmetric PCR and Unlabeled Oligonucleotide Probes

Scott O. Sundberg[1], Jenny Greer[2], Carl T. Wittwer[3] and Bruce K. Gale[2]
1. Bioengineering Dept., 2. Mechanical Engineering Dept. and 3. Pathology Dept.
University of Utah, Salt Lake City, UT

Abstract

Solution-phase, homogeneous DNA melting analysis was performed in 100 nl volumes on a custom microchip/instrument and compared to the gold-standard 10 µl HR-1™ melting instrument for the *VKORC1* C1173T, *CYP2C9*2* and *CYP2C9*3* targets, the genetic targets involved in determining warfarin dosing used in blood thinners. Normal sample was distinguished from heterozygous and homozygous single-nucleotide polymorphisms (SNPs) simultaneously in an array melting platform using an automatic genotype calling melting analysis software.

Introduction

Warfarin is one of the most widely used coumadin anticoagulants but has a wide interindividual dosage variation. Variants in the *VKORC1* and *CYP2C9* genes and certain environmental conditions (i.e. age, bodyweight and drug interactions) account for about 50-60% of the variation in warfarin dosage (Sconce et al. 2005 and Wadelius et al. 2007). Warfarin functions by interfering with vitamin K recycling in the liver. *VKORC1* encodes for the main protein in the vitamin K epoxide reductase complex. Warfarin is metabolized by cytochrome P450 2C9 (*CYP2C9*). Thus, *CYP2C9* *2 and *3 alleles require lower warfarin doses and may have a greater risk of bleeding (Sanderson et al. 2005).

DNA melting analysis as a complement to PCR was introduced in 1997 (Ririe et al. 1997). A dye is included in the PCR that fluoresces in the presence of double-stranded DNA, but not single-stranded DNA. After amplification, fluorescence is monitored as the double-stranded DNA product is slowly heated. When the double helix melts, fluorescence rapidly decreases. The negative first derivative of fluorescence with respect to temperature shows the melting temperature (T_m) as maxima. Recent advances in melting instrumentation (Herrmann et al. 2006) and saturating DNA dyes (Wittwer et al. 2003) allow detection of single nucleotide polymorphisms (SNPs). If the change is heterozygous, DNA heteroduplexes alter the shape of the melting curve (Reed et al. 2004). If the change is homozygous, the absolute temperature of the melting transition shifts (Liew et al. 2004). DNA melting analysis, when compared to existing PCR analytical techniques, is advantageous because it is less complicated, faster (less than 20 minutes for PCR and analysis), and prevents contamination of the sample and environment due to its "closed-tube" format (Zhou et al. 2004). The specific dye used determines the capabilities of the method; LCGreen® Plus detects homozygous and heterozygous sequences well and does not inhibit PCR (Wittwer et al. 2003).

Analysis of the melting transition is often sufficient for genotyping. However, unlabeled probes combined with asymmetric PCR provide even greater specificity over a smaller region, which may be necessary for variant discrimination (Zhou et al. 2005) and is commonly believed to be essential for clinical assays. A recent publication has demonstrated that DNA melting analysis is capable of being miniaturized, thus providing initial results for a microarray chip platform, reducing reagent costs (Sundberg et al. 2007).

Materials and Methods

Microchip Fabrication

Xurography (Bartholomeusz et al. 2005) was used to create a PDMS soft lithography mold, with analysis wells being 1.125 mm in diameter and 0.1 mm in height, creating a 100 nl volume. PDMS was then molded and cured in an oven. PDMS ports were cored and then bonded to glass slide substrates using corona discharge, Figure 1.

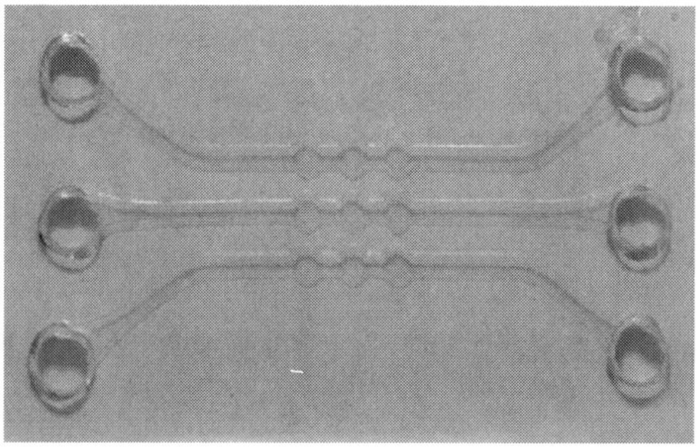

Figure 1. Photograph of the custom microchip showing three channels with three wells per channel. Each well has a diameter of 1.125 mm and a depth of 0.1 mm.

Instrumentation

A Peltier heater (HT4-6-21x43, Melcor, Trenton, NJ) and J-type thermocouple were used for temperature control. Detection used a CCD camera (iXon, Andor Technology, South Windsor, CT) cooled to -80 °C. LEDs (440-460 nm, LXHL-BR02, LUXEON, Philips Lumileds Lighting Company, San Jose, CA) were filtered by an excitation bandpass filter (426-450 nm, Brightline Fluorescent Filters, Semrock, Rochester, New York). Emitted light was collected by a lens (EFS 60mm f/2.8 Macro, Canon, Lake Success, New York) and passed through an emission bandpass filter (467-499 nm, Brightline Fluorescent Filters, Semrock) (Figure 2). All hardware was controlled using LabView 8.0 (NI, Austin, TX).

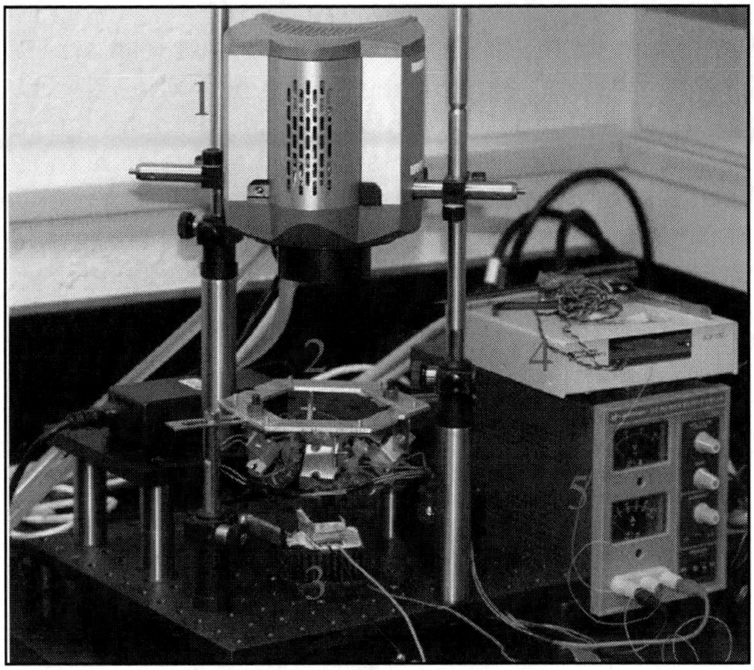

Figure 2. A photograph of the prototype detection instrumentation is shown. (1) CCD camera; (2) LED ring for fluorescence excitation; (3) heating platform using a Peltier heater; (4) thermocouple control; and (5) LED power supply.

Sample Preparation

10X asymmetric PCR with unlabeled oligonucleotide probes was performed on the DNA sample using a LightCycler® and then transferred to an HR-1™ instrument for a high-resolution reference melting curve. 2 μl of the amplified sample was pipetted into the microchip for nanoliter melting analysis. All three genotypes were melted simultaneously and each genotype had three wells for repeat analysis.

Software

The automatic calling software uses an agglomerative, unbiased hierarchical clustering of melting curves (Duda et al. 2001) to make the calls. Given n curves, the two curves that are closest to each other are first determined. The distance between a pair of curves is taken as the mean absolute value of the fluorescence differences between the curves over all temperature acquisitions. The closest two curves are deleted and replaced by their mean, resulting in a new set of n-1 curves. The next nearest pair is then replaced by the weighted mean of that pair. At each step, the weight is the number of original curves that make up each branch being averaged. This process is performed a total of n-1 times until the last pair of curves is merged, producing a dendrogram showing the most likely clusters at each level. The process does not determine the number of clusters (i.e. the number of genotypes) represented by the n curves but does confirm appropriate clustering of samples at each dendrogram level. For these tests three clusters were chosen to represent each of the three genotypes.

Results and Conclusion

Figures 3, 4 and 5 show negative derivative melt plots of normal, heterozygous, and homozygous SNPs within a 190-bp fragment of *VKORC1* C1173T, a 122 bp fragment of

*CYP2C9*2* and 134 bp fragment of *CYP2C9*3*. The 100 nl melts are in good agreement with the 10 μl HR-1 melting curves and each of the genotypes were clearly distinguished in their first melt transition, the unlabeled probe melt region, using an automatic calling software (Idaho Technology) to distinguish the genotypes by setting the cluster number to three. This new platform provides accurate SNP genotyping capability using a hundredth of the volume currently used with gold-standard instrumentation. This technology with further development will lead to a more accurate warfarin dosing within the clinical setting.

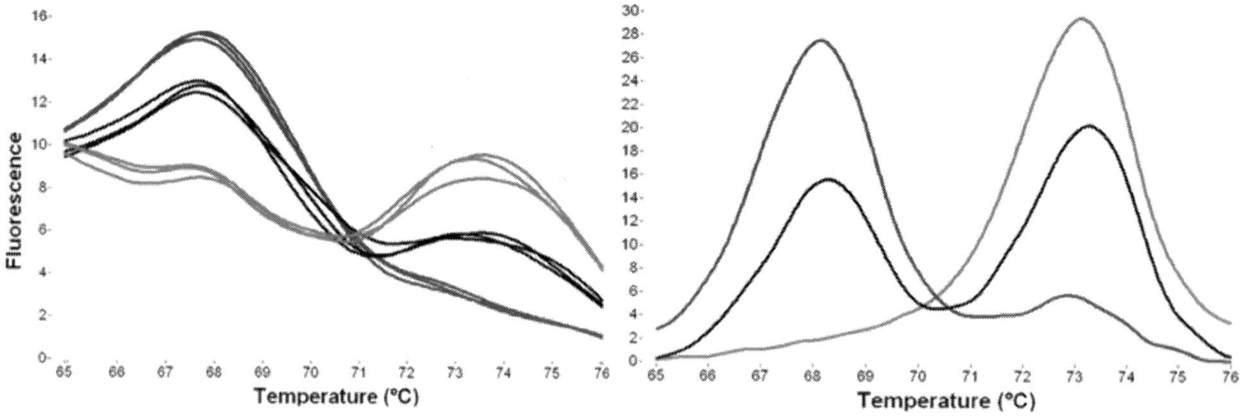

Figure 3. *VKORC1* C1173T negative derivative plots of 100 nl melts (left) and 10 μl melts (right). Normal DNA (red); heterozygous mutation (black); and homozygous mutation (blue).

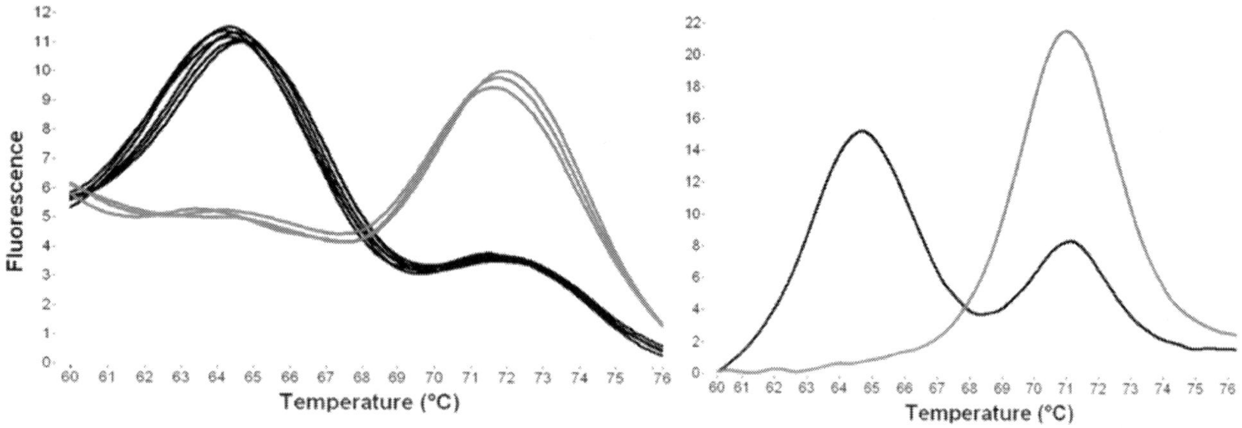

Figure 4. *CYP2C9*2* negative derivative plots of 100 nl melts (left) and 10 μl melts (right). Normal DNA (red) and heterozygous mutation (black).

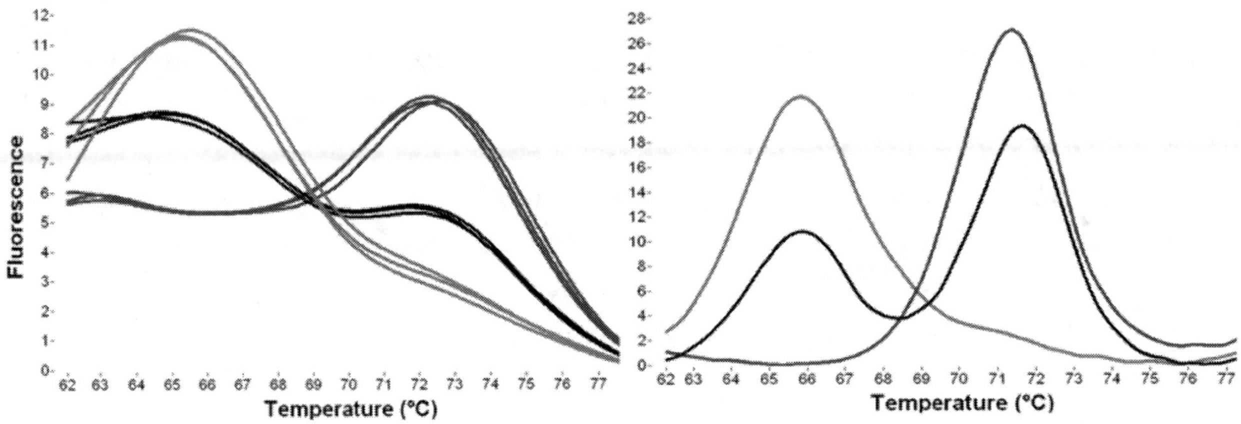

Figure 5. *CYP2C9*3* negative derivative plots of 100 nl melts (left) and 10 μl melts (right). Normal DNA (red); heterozygous mutation (black); and homozygous mutation (blue).

References

Bartholomeusz D, Boutte R and Andrade J. Xurography: rapid prototyping of microstructures using a cutting plotter. *JMEMS* **2005**, *14*:1364-1374.

Duda RO, Hart PE and Stork DG. Pattern classification, 2nd ed. New York: Wiley, 2001:654pp.

Herrmann MG, Durtschi JD, Bromley LK, Wittwer CT and Voelkerding KV. Amplicon DNA melting analysis for mutation scanning and genotyping: cross-platform comparison of instrument and dyes. *Clin Chem* **2006**, *52(3)*:494-503.

Liew M, Pryor R, Palais R, Meadows C, Erali M, Lyon E and Wittwer C. Genotyping of single-nucleotide polymorphisms by high-resolution melting of small amplicons. *Clin Chem* **2004**, *50(7)*:1156-1164.

Reed GH and Wittwer CT. Sensitivity and specificity of single-nucleotide polymorphism scanning by high-resolution melting analysis. *Clin Chem* **2004**, *50(10)*:1748-1754.

Ririe KM, Rasmussen RP and Wittwer CT. Product differentiation by analysis of DNA melting curves during the polymerase chain reaction. *Anal Biochem* **1997**, *245*:154-160.

Sanderson S, Emery J and Higgins J. CYP2C9 gene variants, drug dose, and bleeding risk in warfarin-treated patients: a HuGEnet systematic review and meta-analysis. *Genet Med* **2005**, *7*:97–104.

Sconce EA, Khan TI, Wynne HA, Avery P, Monkhouse L, King BP, Wood P, Kesteven P, Daly AK, and Kamali F. The impact of CYP2C9 and VKORC1 genetic polymorphism and patient characteristics upon warfarin dose requirements: proposal for a new dosing regimen. *Blood* **2005**, *106*:2329–2333.

Sundberg SO, Wittwer CT, Greer J, Pryor RJ, Elenitoba-Johnson O and Gale BK. Solution-phase DNA mutation scanning and SNP genotyping by nanoliter melting analysis. *Biomed Microdevices* **2007**, *9*:159-166.

Wadelius M, Chen LY, Eriksson N, Bumpstead S, Ghori J, Wadelius C, Bentley D, McGinnis R and Deloukas P. Association of warfarin dose with genes involved in its action and metabolism. *Hum Genet* **2007**, *121*:23-34.

Wittwer CT, Reed GH, Gundry CN, Vandersteen JG and Pryor RJ. High-resolution genotyping by amplicon melting analysis using LCGreen. *Clin Chem* **2003**, *49(6)*: 853-860.

Zhou L, Myers AN, Vandersteen JG, Wang L and Wittwer CT. Closed-tube genotyping with unlabeled oligonucleotide probes and a saturating DNA dye. *Clin Chem* **2004**, *50(8)*: 1328-1335.

Zhou L, Wang L, Palais R, Pryor R and Wittwer CT. High-resolution DNA melting analysis for simultaneous mutation scanning and genotyping in solution. *Clin Chem* **2005**, *51(10)*:1770-1777.

Single Molecule λ-DNA Stretching Studied by Microfluidics and Single Particle Tracking

Jun Wang and Chang Lu[a)]

Department of Agricultural and Biological Engineering, School of Chemical Engineering, Birck Nanotechnology Center, Bindley Bioscience Center, Purdue University, West Lafayette

I. INTRODUCTION

New single-molecule manipulation techniques have allowed the studies of the mechanical properties of individual macromolecules. A number of tools, such as optical tweezers[1, 2], atomic force microcope[3, 4], magnetic tweezers[5, 6] and flow field [7, 8], have been applied to exert mechanical forces (~pN) on single molecules and study the forces involved in single molecule biological events.

The mechanical properties of DNA have been intensively studied [2, 3, 6, 9, 10]. The use of optical tweezers led to discovering the striking DNA overstretching transition from B form to S form [2, 9]. At the force of ~65 pN, the double-stranded DNA (dsDNA) extended to 1.7~1.8 times the length of normal B form dsDNA. The highly cooperative overstretching transition plateau has only a narrow band of a few pN. Above that, the stretching profile of dsDNA converges with that of single-stranded DNA (ssDNA). The flexibility and elasticity of dsDNA and ssDNA can be characterized by worm-like chains (WLC) and freely joint chains (FJC), respectively [2, 6, 11]. There are important reasons that DNA stretching is a topic of a great deal of interest. The new states created from DNA stretching are biologically significant since some of them are relevant to DNA-protein interactions [12]. The S-DNA was speculated to provide easier access to the base pairs for transcription purposes [13].

The worm-like chain (WLC) model has been applied to describe the elastic behavior of dsDNA under lower stretching forces [11, 14]. The below formula (1) describes the relationship between the extension x of a WLC and the exerted stretching force F when the dsDNA is entropically stretched ($x \leq 16$ µm for λ DNA).

$$\frac{FP}{k_B T} = \frac{1}{4}(1 - \frac{x}{L_0})^{-2} - \frac{1}{4} + \frac{x}{L_0} \qquad (1)$$

Where k_B is the Boltzmann constant, T is the temperature, P is the persistence length, x is the DNA's extension and L_0 is the contour length. The stretching behavior of DNA is assumed to be inextensible in this equation. An different equation (2) describes the enthalpical stretching of DNA in its extensible WLC regime ($14 \leq x \leq 17$ µm for λ DNA) when x approaches the contour length L_0 defined by B-DNA geometry [15].

$$\frac{x}{L_0} = 1 - \frac{1}{2}(\frac{k_B T}{FP})^{1/2} + \frac{F}{S} \qquad (2)$$

where S is the elastic stretch modulus. At extensions beyond the two regimes, DNA undergoes overstretching.

Flow field has been explored as a tool for exerting pN force on single molecules due to its simplicity and low requirement on instrumentation [6-8, 16]. Hydrodynamic flow fields can generate forces within a very wide range (0.1-1000 pN). This tool potentially offers easy setup

a) Author to whom correspondence should be addressed; email: changlu@purdue.edu and high throughput for observing a number of single molecules simultaneously. In this study, we explore applying microfluidic channels to DNA stretching studies. Compared to homemade channels or chambers applied in previous work, microfabricated channels offer high fidelity and reproducibility in their dimensions and these characteristics are very important for the reproducibility of the inside flow field. Furthermore, due to the high level of integration and automation allowed on a microfluidic platform, chemical and biological assays can be potentially coupled with single molecule force measurement. By tethering one end of a λ-DNA to the glass substrate and the other end to a fluorescent microbead, a number of DNA molecules were attached to the bottom of a microfluidic channel made of PDMS. We changed the flow rate in the microfluidic channel to adjust the hydrodynamic force on the DNA molecules while monitoring the change in the localization of the fluorescent microbead. Using single particle tracking based on fluorescence microscopy we were able to localize the microbead's position with nanometer precision and therefore obtain information about the length of the DNA [17, 18]. We envision that this simple method will become a routine tool for studies related to biomolecule stretching and single molecule force measurement. In addition, we also observed DNA force-extension plateau occurring at low forces (15~30 pN) instead of the commonly reported ~65 pN [2] or 110 pN for unnicked and unwound DNA [19] when the force changing took place at the time scale of seconds.

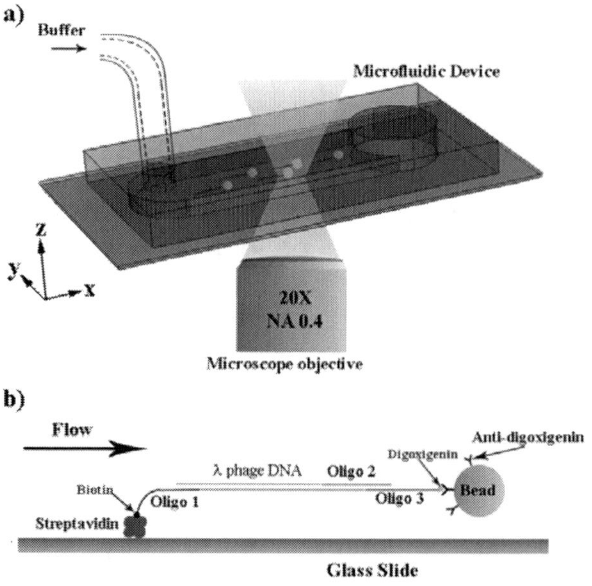

FIG. 1. Schematic of the experimental setup and the bead-DNA-substrate construction. (a) The experimental setup (not drawn to scale). The microfluidic device was mounted on an inverted fluorescent microscope with a 20X objective. The hydrodynamic force was used to pull on the bead. (b) Schematic representation of the bead-DNA-substrate construct. The λ DNA was annealed and ligated to the cohesive ends with Oligo 1 (one end tagged with biotin), Oligo 2 and Oligo 3 (one end tagged with digoxigenin). The λ DNA molecule was bound to glass surface at one end by biotin/streptavidin link and the other end to a fluorescent bead (5 μm diameter) by digoxigenin/anti-digoxigenin link as illustrated.

II. EXPERIMENT

A. Microfluidic Device Fabrication The general information about PDMS microfluidic chip fabrication based on standard soft lithography was provided in our previous publication[20]. The dimensions of fabricated microfluidic channels were 64 μm in the depth, 314 μm in the width and 25 mm in the length.

B. DNA preparation and tethering λ phage DNA (New England Biolabs) with 48,502 base pairs is hybridized to have biotin and digoxigenin in different ends with 3 oligonucleotides (Oligos Etc.). A 3' biotin-modified oligo 1 (5'-Agg TCg CCg AAA AAA AAA AAA AAA AAA AA-biotin-3') was annealed to a 5' overhang of λ phage DNA, while unmodified oligonucleotide 2 (5'-ggg Cgg CgA CCT ggA CAg CAA gTT ggA CCA-3') grafted λ phage DNA with oligonucleotide 3 (5'-digoxigenin-AA AAA AAA AAA AAA Atg gTC CAA CTT gCT gTC C-3') [Fig. 1 (b)]. The resulting dsDNA with biotin and digoxigenin ends has 48,555 base pairs. The dsDNA construct can be denatured to form ssDNA construct by heating in 2mM NaOH at 100 °C for 5 min[21].

200 µl carboxylate-modified 5 µm beads with a concentration of 0.5% W/V (Spherotech Inc.) were centrifuged and resuspended in 50 mM 2-(N-morpholino)ethanesulfonic acid (MES; Sigma-Aldrich) buffer at pH 6.0 for activation. Then the beads were functionalized in 200 µl EDTA-carbodiimide (EDC; Pierce) at 5 mg/ml for 15 minutes and washed in 0.2 M borate buffer at pH 8.5. After incubation for 4 hours with 100 µl anti-digoxigenin(Roche Diagnostics) solution at 0.01 mg/ml, the reaction was quenched by washing the beads with 50 mM Tris solution and then stored in PBS (pH 7.4) with 0.5% BSA and 0.1% Tween-20.

2% 3-aminopropyl-triethoxysilane (APTS; Pierce) in dry acetone was infused into the microfluidic device and incubated for 5 minutes at room temperature. The solution was then removed by deionized water and the channel was cured at 110°C for 30 minutes. After a mixture of 1×10^{-4} mg/ml Streptavidin (Sigma-Aldrich) and 10mg/ml EDC was infused to the channel and incubated for 10 minutes, the reaction was quenched by flowing 50mM NH_4Cl solution in PBS through the channel. Then the channel was filled with blocking buffer (10mM Tris, 2mM EDTA, 10mM NaCl, 0.1% Tween-20, 3mM NaN3, 3mg/ml BSA, pH8.0) for 1 hour. The biotin and digoxigenin modified dsDNA or ssDNA was diluted to 10^{-10} M in DNA bonding solution (50mM Tris, 50mM NaCl, 1mM EDTA, 10% blocking buffer, pH 7.5) before infusing into the channel to allow biotin at one end of DNA bound to Streptavidin for 30 minutes. The excessive DNA was washed by DNA bonding solution with a flow rate at 10 µl/hr for 10 minutes. Finally functionalized microbeads at 1.46×10^{8}/ml were carefully infused into the channel and incubated for 30 minutes. The unbound beads were flushed by the stretching buffer (137mM NaCl, 2mM KCl, 0.5% BSA and 10 mM phosphate buffer) flowing at 10 µl/hr. BSA at this concentration in the stretching buffer does not affect DNA elasticity based on previous literature[21].

C. DNA Stretching Observation To determine that the microbeads are specifically attached to the glass substrate by a DNA molecule, the buffer was flowed in alternating directions at low flow rates (< 50 µL/hr) and we observed the movement of the microbeads. The microbeads attached to the substrate by DNA molecules were able to go either direction. We varied the stretching force by changing the flow rate of the stretching buffer (137mM NaCl, 2mM KCl, 0.5% BSA and 10 mM phosphate buffer) provided by a syringe pump (PHD infusion pump, Harvard Apparatus) with a constant interval of 3s or 10s. Once the force versus extension curves are obtained, we also fit the data using

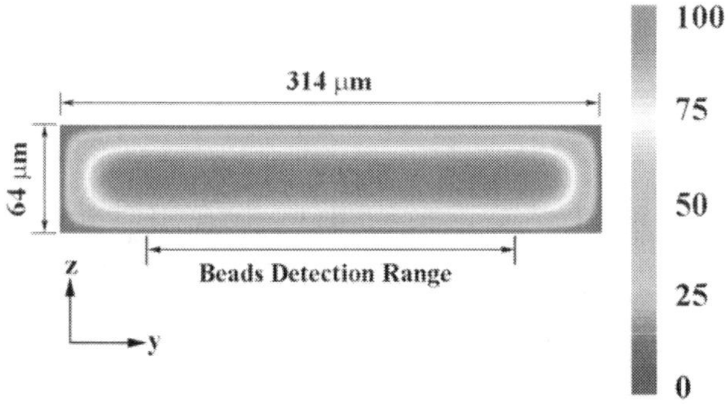

FIG. 2. The steady state velocity profile viewed in the cross-section of the microfluidic channel. The profile was simulated using FLUENT 6. The velocity was normalized by setting 0 at the boundaries and 100 at the center. The region in which tethered beads were monitored is labeled in the figure.

Worm-Like Chain model (WLC) to confirm that single DNA was responsible for the connection between the microbead and the substrate. The velocity profile inside the microfluidic channel was simulated by a commercial finite volume package FLUENT 6 (Fluent Inc.).

The DNA tethered fluorescent microbeads in the microfluidic channel were observed on an inverted microscope (IX-71; Olympus) with a 20X objective. The fluorescent microbeads were illuminated by a 100W mercury lamp filtered by a filter set (Exciter HQ480/40, emitter HQ535/50, and beamsplitter Q505lp; Chroma technology). An area of 433 μm×330 μm was observed by a CCD camera (ORCA-285; Hamamatsu). 50 images were taken at a frame rate of 30 frames/sec for the fluorescent microbeads at a given flow rate for the localization analysis.

D. Nanoscale Localization of the Microbeads The images of fluorescent microbeads are fitted by 2-D Gaussian function defined as

$$P_G(x, y, z_0, A, x_0, y_0, s) = z_0 + A\exp\left(-\frac{(x-x_0)^2+(y-y_0)^2}{2s^2}\right) \tag{3}$$

Where z_0 is constant due to background noise, A is the amplitude, x_0 and y_0 are the bead's fitted fluorescent center, and s is standard deviation of the distribution[17, 18]. The parameters (z_0, A, x_0, y_0, s) are found by nonlinear least-square fitting.

III. RESULTS AND DISCUSION

In this study, microfluidics allows the exertion of pN force on microbeads which are tethered by single-stranded or double-stranded λ-DNA to the substrate. We used fluorescent microbeads which yielded enough signal/noise ratio for us to localize the microbead with nanometer precision (~ 4 nm) at the given moment under a specific hydrodynamic force. We increased the flow rate (the hydrodynamic force) using the syringe pump while recording the localization of the microbead. The length of the DNA can be calculated from the localization of the microbead that the DNA is tethered to. Such data will provide us with the force vs. extension data for single DNA molecules in the image frame.

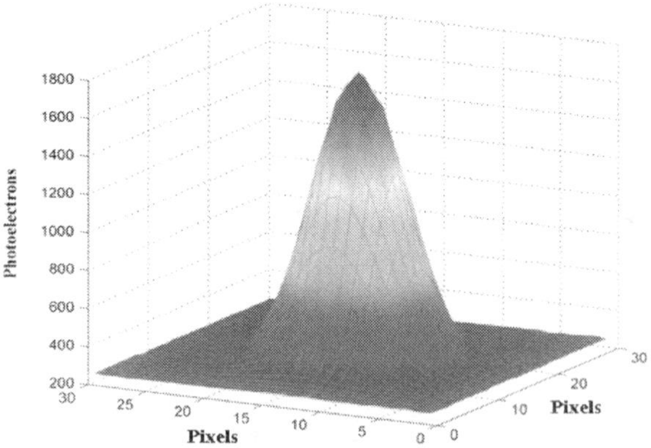

FIG. 3. Point-spread function (PSF) of an individual fluorescent microbead tethered to a single DNA at flow rate 100 μl/hr. A 2-D Gaussian (solid lines) fits the PSF surface very well (R^2=0.997) and this enables the center to be determined with uncertainty within 4.2 nm.

In our experiment, we imaged the DNA tethered fluorescent microbeads in the XY plane. In order to eliminate the effects of the vertical walls on the velocity profile, we only observe the stretching of microbead-DNA-substrate constructs in the center portion of the channel with a distance of at least 50 μm from the two vertical walls. Fig. 2 shows the modeled velocity profile in the cross section of the microfluidic channel. At 50 μm from the vertical walls (XY planes), the no-slip boundaries do not affect the velocity profile any more. The DNA tethered beads being observed experienced essentially the same stretching force at a given flow rate in the channel. Using the protocol we had for DNA tethering to the microbead and the substrate, we typically had 5-8

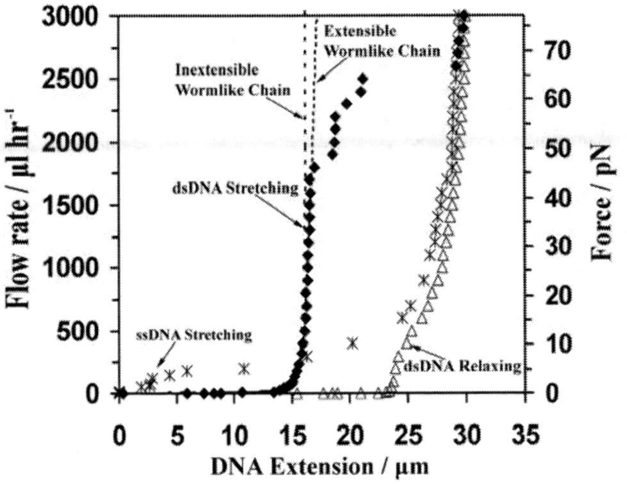

FIG. 4. Flow rate/force vs. extension for dsDNA and ssDNA in a microfluidic channel. The flow rate was increased (stretching) or decreased (relaxing) every 3 seconds. The dsDNA stretching data is compared to a WLC model (dash lines) with persistence length of 53 nm and contour length of 16.5 µm.

DNA tethered beads that were stretched by the hydrodynamic force within the imaging window of 433 µm ×330 µm with a 20X objective. This approach allows us to inspect a small population of single DNA molecules to provide desired statistical significance.

The centers of fluorescent microbeads were determined by two dimensional (2-D) Gaussian fitting of the fluorescent images as described in the experimental section, similar to what is in the literature [17, 18]. Fig. 3 shows that the fluorescent image of a microbead with 20ms exposure time can be fit to a Gaussian with a correlation coefficient of $r^2 = 0.997$. The center, or mean, value of the distribution can be found based on the fitting. The uncertainty σ, or the standard error of the mean in each direction, was 4.2 nm. The fluorescent image allowed us to precisely localize a microbead under a certain flow rate. This process of Gaussian fitting yields precision of several nanometers which is much smaller than the experimental uncertainty caused by Brownian motion, flow irregulaties and mechanical drift [17]. At low flow rates (< 500 µL/hr), Brownian motion of the microbead was significant in the transverse direction which is perpendicular to the flow and parallel to the substrate. We took 50 images within 1.5 seconds and calculated the mean position of the microbead based on all the images.

By gathering the position of the same microbead at different flow rates, we were able to establish the relationship between the flow rate and the DNA extension length. In Fig. 4, we increased the flow rate every 3 seconds between the data points. The force vs. extension curve of a dsDNA molecule matched WLC behavior very well below 1700 µl/hr. When the dsDNA was stretched by over 1700 µl/hr, melting hesteresis happened. At 2600 µl/hr, the DNA was dramatically extended to be 1.77 times of the contour length (~16.5 µm) where it has a plateau in the extension curve. We started to relax the same dsDNA after reaching the flow rate of 3000 µl/hr. The relaxing process was carried out by decreasing the flow rate at the same time interval of 3 sec. The relaxing process displays hysteresis and the extension curve of the relaxed dsDNA closely resembles that of ssDNA stretching with flow rates higher than 500 µl/hr. This suggests that the hysteresis was likely related to the ss nicks presence in the dsDNA. These facts agree with what was described in early literature[2].

Next we calibrate the relationship between the flow rate and the hydrodynamic force exerted on the microbead. In our microfluidic channel, we assume that the hydrodynamic force on the microbeads is proportional to the volumetric flow rate as suggested in the literature (F = ωV) [8]. We use equations 1 and 2 to fit the dsDNA stretching data in Fig. 4 at $x \leq 16$ µm and $14 \leq x \leq 17$ µm, respectively. The parameters involved in the two equations (P, L_0, and S) and the conversion factor ω between the hydrodynamic force F and the volumetric flow rate V were then obtained based on the fitting using least squares method in a Matlab program. We found the persistence length P at 52.88 nm, the contour length L_0 at 16.52 µm, and the elastic stretch modulus S at 1278.58 pN. The conversion factor ω was $0.0257\,\mathrm{pN \cdot hr/\mu l}$ with the particular microchannel applied in the experiment. We could then calibrate the hydrodynamic force at different flow rates [Fig. 4]. The persistence length *P* and the contour length L_0 we obtained from the fitting were consistent with the values in the literature.[6, 14]

Fig. 5 shows that when the force$^{-1/2}$ is plotted against extension based on the dsDNA stretching data presented in Fig. 4, there is a strong linear relationship between the two parameters in the 14-16 μm range. Such linearity is a clear signature of WLC elasticity[11, 22]. The possible interaction between the substrate surface and the DNA-bead, if existent, does not seem to affect the WLC elasticity. It needs to be noted that WLC model is an approximation to pure DNA elasticity due to coupled hydrodynamic drag on DNA molecule even in uniform flow[23]. In our settings, shear flow may also play a role in determining DNA stretching behavior. WLC model was also applied to describe DNA stretching in a shear flow (when DNA was tethered to a surface at one end and free at the other end)[24]. Although it is still different from the case of DNA attaching to a bead at one end (as in our study), this is potentially relevant to the fact the overall stretching we observe here matches the WLC model.

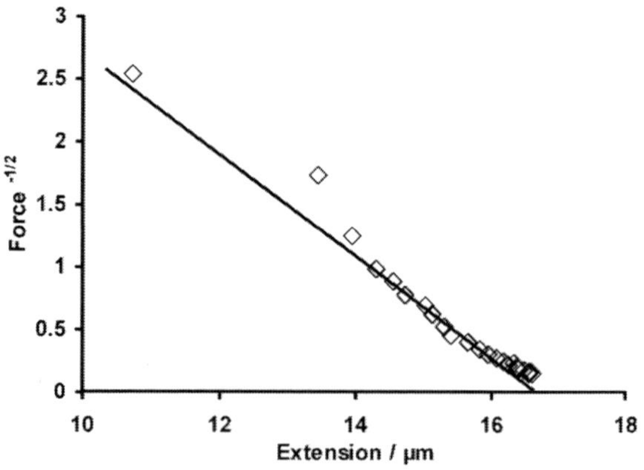

FIG. 5. Force$^{-1/2}$ vs. extension for a single dsDNA molecule. The linear relation between the two in the region of 14-16 μm supports strong stretching WLC behavior. Only the data between 14 and 16 μm was used to generate the trend line.

In Fig. 6(a), we show force vs extension curves of dsDNA taken by varying the flow rate every 10 seconds (totally ~5 min from 100 to 3000 μl/hr). The average loading rate is calibrated to be as low as 0.257 pN/s. These curves can be fitted using equation 1 with Pearson correlation coefficients better than 0.9 (95% or higher confidence) when the force was lower than 15 pN as shown in Fig. 6(b). These curves exhibit multiple transitions and force vs. extension plateaus at low forces in the range of 15-30 pN, compared to ~65 pN reported when the force was varied at the rate

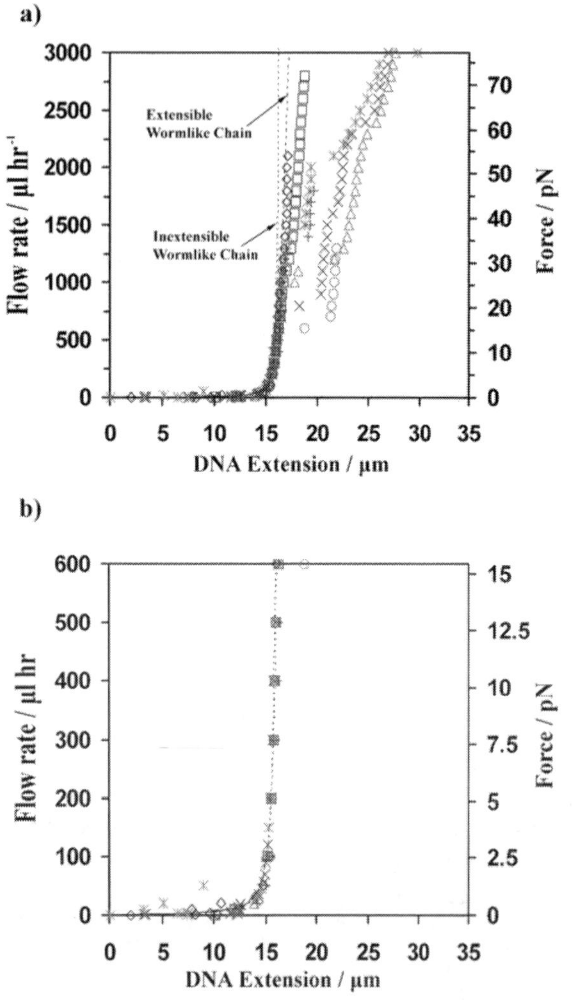

FIG. 6. Flow rate/force vs. extension for a number of single dsDNA molecules in a microfluidic channel. The flow rate was increased every 10 seconds. (a) The force vs. extension curves of several single dsDNA molecules. (b) These curves compare very well to WLC model (dash lines) when the hydrodynamic force is lower than 15 pN.

of 60 pN/μm using optical tweezer[25]. There can be a couple of possible reasons for such behavior. First, the force plateau has been suggested to be associated with either DNA conformational transitions [9, 26] or DNA melting [27]. In our case, the possible mechanism for the plateaus at low forces is due to the force displacement over large time scale (force changed every 10 seconds). It is known from ligand-receptor pair that its unbinding force depends on both the molecular fundamental property and the loading rate [28]. The unbinding force is smaller if the load on the complex increases at a sufficiently slow rate, because there is more time for thermal fluctuations to drive the system over the energy barrier. The low force needed for dsDNA stretching plateua is consistent with the prediction that low force with long duration can accelerate thermal activation over a free energy barrier stabilizing dsDNA[29]. Such slow stretching processes are likely to be involved in real biomolecule functioning for them to work efficiently to reduce unnecessarily energy dissipation through molecular friction. Second, strictly speaking, the DNA-bead system experiences the extensional force generated by the flow-bead interaction and in the meantime it is also sheared in Poiseuille flow between the channel floor and ceiling. It is possible that the shear flow can influence the DNA stretching in a more significant way when the loading rate is low and the flow rate is high. Furthermore, the interaction between the DNA-bead with the substrate surface can vary with different flow rates. These factors could potentially contribute to the plateau change.

IV. CONCLUSION

We validate the application of microfluidics and single particle tracking for the study of DNA stretching. Microfabrication provides devices with highly reproducible dimensions that are ideal for establishing flow field for single DNA molecule stretching. The DNA length is obtained by localizing the fluorescent microbead attaching to the one end of a single DNA molecule (the other end tethered to the substrate). The high signal to noise ratio of the fluorescent microbead allows nanometer localization with a common CCD camera. We found the multiple transitions and plateau taking place in the stretching curve. The study of the DNA stretching based on microfluidics provides a versatile platform for incorporating chemical and biological assays during force measurement. Further theoretical and experimental work will need to be carried out to precisely characterize the forces and interactions experienced by DNA molecules in the settings of our method.

ACKNOWLEDGEMENT

The authors thank Dr. Chengde Mao at Purdue University and Dr. Min Wu at Yale University for helpful discussions about the experimental procedures. The research was supported by Purdue University.

REFERENCE

1 Wang, M. D., Yin, H., Landick, R., Gelles, J. and Block, S. M. (1997) Stretching DNA with optical tweezers. *Biophysical Journal*, **72**, 1335-1346

2 Smith, S. B., Cui, Y. J. and Bustamante, C. (1996) Overstretching B-DNA: The elastic response of individual double-stranded and single-stranded DNA molecules. *Science* ,**271**, 795-799

3 Rief, M., Clausen-Schaumann, H. and Gaub, H. E. (1999) Sequence-dependent mechanics of single DNA molecules. *Nature Structural Biology,* **6**, 346-349

4 Rief, M., Gautel, M., Oesterhelt, F., Fernandez, J. M. and Gaub, H. E. (1997) Reversible unfolding of individual titin immunoglobulin domains by AFM. *Science,* **276**, 1109-1112

5 Strick, T. R., Allemand, J. F., Bensimon, D., Bensimon, A. and Croquette, V. (1996) The elasticity of a single supercoiled DNA molecule. *Science*, **271**, 1835-1837

6 Smith, S. B., Finzi, L. and Bustamante, C. (1992) Direct Mechanical Measurements of the Elasticity of Single DNA-Molecules by Using Magnetic Beads. *Science*, **258**, 1122-1126

7 Lee, J. B., Hite, R. K., Hamdan, S. M., Xie, X. S., Richardson, C. C. and van Oijen, A. M. (2006) DNA primase acts as a molecular brake in DNA replication. *Nature*, **439**, 621-624

8 van Oijen, A. M., Blainey, P. C., Crampton, D. J., Richardson, C. C., Ellenberger, T. and Xie, X. S. (2003) Single-molecule kinetics of lambda exonuclease reveal base dependence and dynamic disorder. *Science*, **301**, 1235-1238

9 Cluzel, P., Lebrun, A., Heller, C., Lavery, R., Viovy, J. L., Chatenay, D. and Caron, F. (1996) DNA: An extensible molecule. *Science*, **271**, 792-794

10 Clausen-Schaumann, H., Rief, M., Tolksdorf, C. and Gaub, H. E. (2000) Mechanical stability of single DNA molecules. *Biophysical Journal*, **78**, 1997-2007

11 Marko, J. F. and Siggia, E. D. (1995) Stretching DNA. *Macromolecules*, **28**, 8759-8770

12 Leger, J. F., Robert, J., Bourdieu, L., Chatenay, D. and Marko, J. F. (1998) RecA binding to a single double-stranded DNA molecule: a possible role of DNA conformational fluctuations. *Proceedings of the National Academy of Sciences of the United States of America,* **95**, 12295-12299

13 Ahsan, A., Rudnick, J. and Bruinsma, R. (1998) Elasticity theory of the B-DNA to S-DNA transition. *Biophysical Journal*, **74**, 132-137

14 Bustamante, C., Marko, J. F., Siggia, E. D. and Smith, S. (1994) Entropic elasticity of lambda-phage DNA. *Science*, **265**, 1599-1600

15 Odijk, T. (1995) Stiff Chains and Filaments under Tension. *Macromolecules*, **28**, 7016-7018

16 Davenport, R., Wuite, G., Landick, R. and Bustamante, C. (2000) Single-molecule study of transcriptional pausing and arrest by E. coli RNA polymerase. *Science*, **287**, 2497-2500

17 Thompson, R. E., Larson, D. R. and Webb, W. W. (2002) Precise nanometer localization analysis for individual fluorescent probes. *Biophysical Journal*, **82**, 2775-2783

18 Yildiz, A., Forkey, J. N., McKinney, S. A., Ha, T., Goldman, Y. E. and Selvin, P. R. (2003) Myosin V walks hand-over-hand: Single fluorophore imaging with 1.5-nm localization. *Science*, **300**, 2061-2065

19 Léger, J. F., Romano, G., Sarkar, A., Robert, J., Bourdieu, L., Chatenay, D. and Marko, J. F. (1999) Structural Transitions of a Twisted and Stretched DNA Molecule. *Physical Review Letters*, **83**, 1066-1069

20 Wang, J. and Lu, C. (2006) Microfluidic cell fusion under continuous direct current voltage. *Applied Physics Letters*, **89**, -

21 Maier, B., Bensimon, D. and Croquette, V. (2000) Replication by a single DNA polymerase of a stretched single-stranded DNA. *Proceedings of the National Academy of Sciences of the United States of America***,** **97**, 12002-12007

22 Baumann, C. G., Smith, S. B., Bloomfield, V. A. and Bustamante, C. (1997) Ionic effects on the elasticity of single DNA molecules. *Proceedings of the National Academy of Sciences of the United States of America*, **94**, 6185-6190
23 Perkins, T. T., Smith, D. E., Larson, R. G. and Chu, S. (1995) Stretching of a Single Tethered Polymer in a Uniform-Flow. *Science*, **268**, 83-87
24 Ladoux, B. and Doyle, P. S. (2000) Stretching tethered DNA chains in shear flow. *Europhysics Letters*, **52**, 511-517
25 Wenner, J. R., Williams, M. C., Rouzina, I. and Bloomfield, V. A. (2002) Salt dependence of the elasticity and overstretching transition of single DNA molecules. *Biophysical Journal*, **82**, 3160-3169
26 Lebrun, A. and Lavery, R. (1996) Modelling extreme stretching of DNA. *Nucleic Acids Research*, **24**, 2260-2267
27 Williams, M. C., Wenner, J. R., Rouzina, I. and Bloomfield, V. A. (2001) Effect of pH on the overstretching transition of double-stranded DNA: evidence of force-induced DNA melting. *Biophysical Journal*, **80**, 874-881
28 Evans, E. and Ritchie, K. (1997) Dynamic strength of molecular adhesion bonds. *Biophysical Journal*, **72**, 1541-1555
29 Strunz, T., Oroszlan, K., Schafer, R. and Guntherodt, H. J. (1999) Dynamic force spectroscopy of single DNA molecules. *Proceedings of the National Academy of Sciences of the United States of America*, **96**, 11277-11282

119b Chemotaxis of E. Coli in a Microfluidic Device: Experiments and Simulations

Rajitha Vuppula, Chemical Engineering, IIT Bombay, Powai, Mumbai - 400 076, India, Mahesh S. Tirumkudulu, Department Of Chemical Engineering, IIT Bombay, Powai, Mumbai - 400 076, India, and K. V. Venkatesh, Chemical Engineering, Indian Institute of Technology Bombay, Dept of Chemical Eng, IIT Bombay Powai, Mumbai, 400076, India.

Chemotaxis is the phenomena by which microorganisms find food, or move away from poisonous chemicals. Chemicals which are capable of eliciting such a response from chemotactic cells are called chemotactants. Chemotactants may act as attractants (chemoattractants), in which case the chemotactic cell will move towards them, thereby exhibiting a positive response. Some chemicals act as repellents (chemorepellents), in which case the chemotactic cell will move away from them, thereby exhibiting a negative response. The understanding of this behavior is important, because of its relevance in cancer, immunology and microbiology.

In the present work, we describe a simple method to create stable chemical gradients in a microfluidic device in the absence of fluid flow. The microfluidic device is fabricated in poly(dimethylsiloxane) (PDMS) using soft-lithography. The gradients are measured in both, space and time using a fluorescent glucose 2-(N-(7-nitrobenz-2-oxa-1,3- diazol-4-yl)amino)-2-deoxyglucose (2-NBDG) which acts as chemoattractant for E.coli. We quantify the chemotaxis of E.coli to different gradients and compare the experimental results with the predictions of existing theoretical models of the signaling pathways.

119c Microfluidic Devices For Local Rheological Characterization Of Viscoelastic Biological Materials

Danial N. Hohne, Chemical Engineering, University of Michigan, 3074 HH Dow, 2300 Hayward St., Ann Arbor, MI 48109, John G. Younger, Department of Emergency Medicine, University of Michigan, 4063 BSRB Box 2200, Ann Arbor, MI 48109, and Michael J. Solomon, Department of Chemical Engineering, University of Michigan, 2300 Hayward Street, Ann Arbor, MI 48109.

Stahphylococcus epidermidis and klebsiella pneumoniae are examples of bacteria that produce an extensive network of polysaccharides that lead to formation of a biofilm as an exterior capsule to the cellular pathogen. This extracellular matrix helps to create a protective environment in which the bacteria can proliferate. The extracellular structure of biofilm aggregates can anchor to biomedical devices, such as catheters, leading to bloodstream infection. The shedding of these aggregates into the blood stream and their behavior and fate in the circulatory system is hypothetically linked to the rheological response of the capsular aggregates. However, mechanical rheology measurements on these materials are challenging because of sample volume requirements and material heterogeneity. Thus, we have produced a microfluidic device to determine the rheological response of biofilm aggregates on scales relevant to the circulatory system. The device uses flexible valves fabricated in poly(dimethyl siloxane) according to the methods of Struder and co-workers (Struder V, Hang G, Ortiz M, Anderson WF and Quake SR, J. Appl. Physics 95(1), 393-398, 2004) to deform bacterial aggregates grown at biologically relevant shear rates in the microfluidic environment. We find that the response of the valve membrane is correlated with the elastic properties of the material loaded into the device. This response is measured in a model material, gellan gum, using microrheological particle tracking of polystyrene spheres suspended in the elastic matrix. The particle trajectories are compared with the results of finite element simulations using linearly elastic constitutive models. The approach of comparing between microrheology and finite element simulation is extended to probe the local elastic response of biofilms of varying intra and extracellular structure through genetic modification of the pathogens.

119d Linear Amplification Of Mrna Using Microfluidic Systems For Gene Expression Analysis

Jason G. Kralj[1], Audrey Player[2], Ernest Kawasaki[2], and Laurie E Locascio[1]. (1) Biochemical Sciences, National Institute of Standards and Technology, 100 Bureau Dr, MS 8313, Gaithersburg, MD 20899, (2) National Cancer Institute/National Institutes of Health, 8717 Grovemont Circle, Bethesda, MD 20877

Many researchers are currently pursuing the development of unbiased, accurate methods to measure the whole proteome or the whole transcriptome in small biological samples at or near the single-cell level in order to effectively identify unique biomarkers of disease. In the work described here, whole transcriptome amplification of Jurkat T-cell mRNA was achieved using a 2-step linear amplification method with the first step being the Eberwine amplication method performed in a packed bed on a microfluidic chip. The Eberwine amplification method works without themocycling or bias and gives ~1000-fold increase in concentration in benchtop experiments, though it requires relatively high starting concentrations. For this amplification step, microfluidics offer the great advantage of dramatically increasing the effective concentration of the original starting material through immobilization of the cDNA library onto microbeads packed in a ~1 nL reaction volume. The second amplification step is performed off-chip, and results indicate that we are able to raise the sample concentration enough for use on microarrays.

The reaction was performed in a device made by multilayer soft-lithography techniques with hydraulic valves that close partially to retain the cDNA-functionalized microbeads under constant flow. Specific high and moderate-copy genes were detected post-amplification using off-chip RT-PCR to confirm amplification. RNA length (i.e. amplified sample quality) was measured using capillary electrophoresis and found to match the average template size (1 kb), indicating little or no bias was introduced in the sample that would lead to false gene expression profiling.

119e Label-Free DNA Detection In Microchannels By A Reversible Light Diffracting Mesophase

Faisal Shaikh and Victor M. Ugaz. Artie McFerrin Department of Chemical Engineering, Texas A&M University, 3122 TAMU, College Station, TX 77843

We introduce a new technique to enable label-free detection of charged biomolecules (DNA, proteins) that harnesses a microfluidic on-chip electrode array to generate localized zones where the biomolecules become ultra-concentrated, resulting in formation of a mesophase that is visible under ordinary white light. We illustrate this phenomenon in a series of experiments that involve injecting a dsDNA sample into a microchannel and initiating the focusing and compaction process by applying a 2 V DC potential between a pair of electrodes in the array. The DNA present between the electrode pair migrates toward the anode in response to the electric field and experiences 2-D compaction that confines the biomolecules within a layer at the electrode surface that is only a few microns thick. Consequently, when the electrode is viewed under white light it becomes very bright and is clearly distinguishable. Distinct and colorful textures are also evident when the mesophase is observed under transmitted polarized light using an optically transparent indium tin oxide electrode array. The dramatic magnitude of this effect combined with the extremely low actuation voltages required make this technique a robust label-free method to enable minute sample quantities to be detected in portable bioanalysis systems without the use of chemical fluorophores.

119f Integrated Micro System For High Resolution Imaging And High-Throughput Sorting Of C. Elegans

Kwanghum Chung, Matthew Crane, and **Lu Hang**. Gatech, 311 Ferst Dr. NW, Atlanta, GA 30332

C. elegans is an excellent model system for studying the genetics of neural development and behavior. In vivo microscopy has been extensively used for phenotypical characterizations and visual screens. Despite the long history, the current procedure is manual, low-throughput, and requires anesthetics. We present a high-throughput automated micro system for high-resolution imaging and sorting of worms without the use of anesthetics. As a part of this system, we developed a microfluidic device that route, trap, and immobilize live animals. The device is a multi-layer design: samples are manipulated in the bottom layer. The top layer includes channels for actuating valves and circulating cooling fluid. The design allows high-resolution and high-magnification imaging using standard microscopes. In order to prevent multiple worms from entering the imaging zone, we designed a self-regulated loading scheme. The nematodes small thermal mass allows it to be cooled and stopped in <1sec without anesthetics. As a result, the time required for sorting is less than 2 seconds. We have developed integrated algorithms which fully automate the processes outlined above and allow for high throughput experiments. To demonstrate capability of this system as a high resolution imaging tool we obtained Z-stack images of green fluorescent protein (GFP)-expressing neurons of a stopped worm with a 100X oil objective. Those images show the resolution of single axons/dendrites, comparable to that of current standard optical microscopy. We show a high-speed sorting of worms with or without GFP.

119g Generation Of Microenvironments With Controlled Mechanical Properties For Analysis Of Mouse Embryonic Stem Cell Fate

Bong Geun Chung, **Benjamin S. Gillers**, Shamit Shrivastava, and Ali Khademhosseini. Harvard-MIT, Division of Health Science and Technology, Massachusetts Institute of Technology, Cambridge, MA 02139

The mechanical aspects of the cellular microenvironment have been shown to play an important role in controlling adult stem cell self-renewal and differentiation. However not much has been studied with respect to the effects of these forces on embryonic stem cells (ESCs). Therefore, the development of technology that can be used to study the effects of mechanical forces on ESCs and embryoid bodies (EBs) will be of great importance in directing the differentiation of ESCs. Here we have used poly (ethylene glycol) (PEG) hydrogels to control the mechanical environment of ESCs since the mechanical and physical properties of PEG hydrogels can be controlled without providing biochemical signals to cells. We present a platform to encapsulate ESCs and EBs within PEG hydrogels of controlled geometries, mechanical, and diffusive properties such that the effects of diffusion and mechanics can be independently controlled. Furthermore we have performed preliminary analysis of the effects that these changes on the proliferation and differentiation of encapsulated mouse ESCs and EBs. It was found that cells can be encapsulated either as individual cells or cell aggregates within PEG hydrogels of various molecular weights, concentrations, and dimensions without losing viability for at least 7 days. Current experiments are underway to study the effects of these controlled microenvironments on ESC differentiation and proliferation.

152a Gel-Free Isoelectric Focusing In Plastic Microfluidic Devices

Z. Hugh Fan, Champak Das, and Jiyou Zhang. Mechanical and Aerospace Engineering, University of Florida, PO Box 116250, Gainesville, FL 32611-6250

We will present the investigation on isoelectric focusing (IEF) in plastic microfluidic devices. A laser-induced fluorescence (LIF), whole-channel imaging detection (WCID) system has been developed to monitor proteins while they migrate under an electric field. IEF is carried out in a separation medium consisting of carrier ampholytes and a mixture of linear polymers (hydroxyethyl cellulose and hydroxypropyl cellulose). Dynamic coating of the linear polymers prevents proteins from adsorption and suppresses electroosmotic flows. We experimentally confirmed the theoretical prediction that IEF resolution is essentially independent of the focusing length when the applied voltage is kept the same and within a range that it does not cause Joule heating. This result supports the notion that IEF in a microfabricated device leads to more rapid analysis without sacrificing the resolving power. A higher separation voltage also brings about more rapid analysis and superior separation resolution. We found that a linear relationship exists between the focusing time and the inverse of the electrical field strength. IEF of proteins is achieved in 1.5 minutes when 500 V was applied across a 1.9-cm channel. We are incorporating IEF into a microfluidics-enabled two-dimensional electrophoresis device that has a potential to be used for proteomics research.

152b Quantitative Toxicoproteomic Analysis Of Carcinogen-Treated Animal Tissues And Human Cells For Human Health Risk Assessment

Yue Ge, Julian Preston, and Russell D. Owen. Environmental Carcinogenesis Division, US Environmental Protection Agency, Research Triangle Park, NC 27711

Humans are exposed to a variety of environmental toxicants, and this together with a large number of interacting factors can contribute to an individual's risk for health. To understand the toxic mechanisms and/or modes of action for human health risk assessment, molecular characterization of the proteome is required for a complete understanding of chemical-induced toxic responses. As an integration of proteomics, toxicology, and bioinformatics, toxicoproteomics focuses on proteome changes in cells or tissues following exposure to toxicants and aids in identifying the changes that may be important to disease development. To apply quantitative toxicoproteomic analysis to the identification of toxic mechanisms and modes of action as well as to biomarkers of exposure, toxicity, and effect for human health risk assessment, we have developed an integrated toxicoproteomic platform. This platform has been applied to the analysis of the effects of several environmental carcinogens such as arsenic, bromate, benzo[a]pyrene, and conazoles on protein expression, phosphorylation, and oxidation in mouse and rat liver, lung, thyroid, and kidney and in human cells. Our results demonstrate that toxicoproteomics has the potential to be a powerful tool to detect the "fingerprints" of specific toxic effects, and thereby to facilitates the identification and development of biomarkers for risk assessment.

152c Quantitative Proteomic Analysis Of A Soil Bacterium Under Different Levels Of Cadmium Stress

Carla M. R. Lacerda[1], Phillip C. Wright[2], and Kenneth F. Reardon[1]. (1) Department of Chemical and Biological Engineering, Colorado State University, 1370 Campus Delivery, Fort Collins, CO 80523-1370, (2) Department of Chemical and Process Engineering, The University of Sheffield, Biological & Environmental Systems Group, Sheffield, S1 3JD, United Kingdom

In a previous study, metaproteomics was used to obtain functional information about the response of soil microbial communities to cadmium stress. In that study, we were able to separate and identify a number of proteins that were differentially expressed due to the exposure to toxic levels of cadmium.

In an effort to better understand the roles played by proteins in cadmium resistance mechanisms, we isolated the bacterium most resistant to cadmium. A defined minimal medium was designed to avoid cadmium precipitation, and the microbial consortium was grown in increasing cadmium. During these isolation experiments, only a very small number of distinct colonies were capable of growing in high cadmium-level environments. The bacterium used for this study was selected for its ability to grow in the presence of 100 mg/L cadmium. Its 16S rDNA was sequenced, and the organism was identified as a strain of *Burkholderia cepacia*.

The specific objective of this investigation was to assess the mechanisms used by this bacterium to adapt to high levels of cadmium. To accomplish this, we used stable isotope metabolic labeling, in which either N-14 or N-15 was supplied as the only nitrogen source in the medium. Two cadmium concentrations were used: 10 and 100 mg/L of cadmium. Cultures were harvested at an OD600 of approximately 0.6, and equal amounts of labeled and unlabeled cultures were combined. Proteins were extracted and analyzed by 2DE and shotgun proteomics. The use of metabolic labeling yielded quantitative information about differential protein expression and increased the confidence level of protein identifications by adding an extra constraint that confirmed the presence of up- or down-regulation of two or more peptides per protein, also in agreement with gel images. A number of those proteins were identified from both workflows after database searching and de novo sequencing. There was a clear difference in the proteomes of the cultures grown on high and low level of cadmium, including proteins involved in energy metabolism, defense and transport. This approach represents one of the first applications of stable isotopic labeling to environmental proteomics of soil isolates.

152d Development Of Flamingo™, A Novel Fluorescent Dye For Non-Specific Detection And Quantitation Of Proteins In Gels

Tom Berkelman, Life Science Group, Bio-Rad Laboratories, 6000 James Watson Dr., Hercules, CA 94547

Gel-separated proteins are most commonly detected and quantitated by dye binding, utilizing the property of some dyes to bind to proteins non-specifically, rendering the proteins optically detectable and quantifiable. Fluorescent dyes are widely used for this due to the high sensitivity and dynamic range of fluorescent detection. Flamingo is a novel dye that was developed from a class of dyes that are minimally fluorescent at low pH in the absence of protein, but acquire strong fluorescence in the presence of denatured protein. This property was reasoned to be potentially useful for staining proteins separated in polyacrylamide gels, which are typically denatured and fixed in acid solution. Modifications of the dye structure were undertaken to enhance protein binding, resulting in development of a dye with particularly useful gel staining properties. Fixed protein gels can be stained by immersion in an acid solution of Flamingo. High sensitivity protein detection is possible without the need for destaining, since unbound dye is relatively non-fluorescent. Investigations of Flamingo Fluorescent Gel Stain were undertaken to characterize the product and compare it to other fluorescent stains with respect to gel staining sensitivity, linearity of response to protein, protein-to-protein variability and compatibility with mass spectrometric analysis. Flamingo was found to allow protein detection down to 30 pg and allow linear quantitation over four orders of magnitude. Flamingo staining was found to be fully compatible with peptide mass fingerprinting by MALDI-MS and to result in a lower incidence of oxidative protein modification than SYPRO™ Ruby.

152e Functional Proteomic Approach for Ion-Related Outer Membrane Proteins

Xuanxian Peng[1], Xiangmin Lin[1], Lina Wu[1], Sanying Wang[2], Hui Li[1], and Chuangzhong Huang[1]. (1) State Key Laboratory of Biocontrol, School of Life Sciences, Zhongshan University, Guangzhou, 510275, China, (2) Department of Biology, Xiamen University, Xiamen, 361005, China

Outer membrane (Om) proteins play key role in response to environmental changes in Gram-negative bacteria. Here we introduce the progress on functional proteome responded to iron and Na ions.

Iron is an essential element for bacterial growth. The Om proteins of Gram-negative bacteria play an important role in overcoming the conditions of poor iron availability in the host and natural environments. We undertook a comparative analysis on a sarcosine-insoluble fraction of E. coli K-12 cultured in LB medium with or without limited iron by 2-DE proteomic methodologies. Three proteins, OmpW, OmpX and Tsx, related to iron homeostasis first reported here were further investigated. Importantly, our further proteomic studies reveal a number of previously unrecognized a membrane-interacting multi-protein complex, which may include OmpW, Frd (A, B, C and D), Opd1, Slp and YbjP. Of the eight proteins, OmpW and FrdB are iron-related proteins and bind together. The multi-protein complex is disassociated at 200 ìmol DIP, and OmpW and FrdB mutants show significantly physiological changes.

Bacteria can shift between the ambient seawater and hosts, suggesting the existence of proteins rapidly responding to salt concentrations. In our study, proteomics methodologies were applied to screen OMPs related to salt stress in V. parahaemolyticus, V. alginoticus, P. damsela. OmpW and OmpV were determined in the response in these bacteria as OmpC and OmpF did in E. coli. Furthermore, the two genes were overexpressed in E. coli and complemented in V. paraheamolyticus mutants. The ability in salt-tolerance was elevated in the bacteria overexpressed OmpW and reduced in the cells overexpressed OmpV. V. paraheamolytixus mutants could recover their responses to environmental salt concentrations when they were complemented by OmpW and OmpV of the three cells. These findings indicate that OmpW and OmpV are required for environmental salt regulation in these bacteria.

Blue native PAGE analysis of DLDH

Liang-Jun Yan, Ph.D.
Department of Pharmacology and Neuroscience
University of North Texas Health Science Center at Fort Worth

1 Introduction

Dihydrolipoamide dehydrogenase (DLDH, EC 1.8.1.4) is the third component of pyruvate dehydrogenase, α-ketoglutarate dehydrogenase, and branched chain keto acid dehydrogenase [1,2]. It is also the L protein in the glycine cleavage system [3,4]. The active form of DLDH is a stable homodimer, with each monomer being a 50 kDa protein that owns a non-covalently but tightly bound FAD molecule and a redox active center containing two cysteine residues that are directly engaged in thiol-disulfide redox reactions during catalysis [1,5,6]. *In vivo*, DLDH oxidizes dihydrolipoamide that is covalently linked to acyltransferase using NAD^+ as the electron acceptor, leading to the release of NADH molecule [1,2,7]. *In vitro*, however, DLDH possesses diaphorase activity [8] that is capable of transferring electrons from NADH to a variety of electron acceptors including cytochrome c [9], ubiquinone [10,11], and electron accepting dyes such as 2,6-dichlorophenolindophenol (DCPIP) [12] and nitroblue tetrazolium (NBT) [13,14]. Both the *in vivo* and *in vitro* reactions have been widely used as the bases for measuring DLDH activity [12], but none of those measurements offers the capability of native DLDH isolation from animal tissues and simultaneous DLDH activity visualization and quantification, which could be conducive to further analysis of DLDH by techniques such as mass spectrometry in studying the mechanisms of DLDH post-translational modifications under a variety of experimental conditions.

Blue native polyacrylamide gel electrophoresis (BN-PAGE) is a method originally developed for the separation of mitochondrial electron transport chain complexes [15]. The method involves protein binding of Coomassie brilliant blue (CBB) G-250 before gel loading. As CBB G-250 binding introduces a charge shift on the protein surface, and yet does not denature proteins, so proteins (and protein complexes) separate in the gel according to their surface charges, their shapes and their native molecular masses [16,17]. Additionally, BN-PAGE also introduces the use of aminocaproic acid that improves the solubilization of membrane proteins [15]. BN-PAGE has gained high popularity since its development in the early 1990's and has been successfully used for separation of water soluble proteins [18,19], nuclear protein complexes [20] as well as whole cellular lysates [21]. In the literature, a well-documented application of BN-PAGE is the localization and quantification of mitochondrial NADH ubiquinone oxidoreductase (complex I) activity visualized by NADH and NBT staining [15,22-25]. Recently, comprehensive BN-PAGE protocols and their potential applications have been published [26,27].

Given that DLDH, *in vitro*, is capable of reducing NBT at the expense of NADH [13,14], it should also be possible to devise a gel-based, histochemical staining method for quantification of DLDH diaphorase activity. The purpose of this study was, therefore, to develop a BN-PAGE-based method for isolation of enzymatically active DLDH from animal tissues and visualization as well as quantification of its diaphorase activity using the NADH/NBT detection system. We now report our development of such a method with rat brain mitochondria as the source of DLDH.

2 Results

2.1 DLDH diaphorase activity staining in mitochondrial extracts

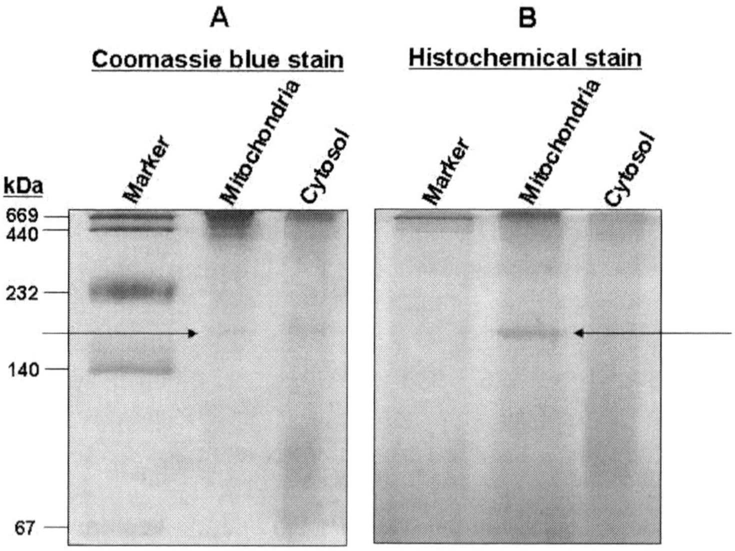

Figure 1. BN-PAGE resolution of mitochondrial extracts and cytosolic proteins. Shown are Coomassie blue staining (A) and activity staining (B) of the gels following BN-PAGE. Ten microgram proteins were loaded in each lane. Mitochondrial extract was prepared as described in the text in a hypotonic buffer containing 20 mM potassium phosphate, pH 7.4, 1% Triton X-100, 2 mM EDTA and 2 mM EGTA. Arrows indicate the DLDH bands on both gel images. Native gel protein markers used in this figure as well as in other figures as indicated are: thyroglobulin, 669 kDa; ferritin, 440 kDa; catalase, 232 kDa; lactate dehydrogenase, 140 kDa; BSA, 66 kDa.

2.2 Confirmation of DLDH identity

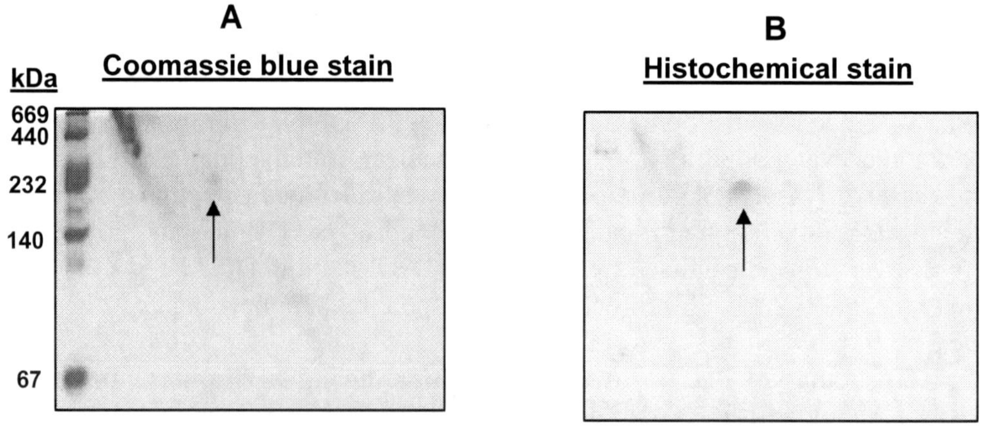

C

LC/ESI-MS/MS generated peptide	*BLAST-Matched peptide*	*Identity*
ALTGGIAHLFK	133 ALTGGIAHLFK 143	Zebrafish DLDH
RRPFTQNLGLEELGIELDPK	316 RPFTQNLGLEELGIELDPK 334	Mouse DLDH
KSEEQLKEEGVEFK	404 KSEEQLKEEGIEFKI 418	Rat DLDH
KALLNNSHYYHLAHGKD	89 KALLNNSHYYHLAHGKD 105	Golden hamster DLDH
RLGADVTAVEFLGHVGGIGIDM*EISKN	234 RLGADVTAVEFLGHVGGIGIDM-EISKN 260	Chinese hamster DLDH

Figure 2. Mass spectral peptide sequencing confirming that the protein of interest in Fig. 1 was indeed DLDH. Following BN-PAGE, gel strips without activity staining were horizontally placed immediately onto the second dimensional native gels (7.5% resolving gel). Shown are Coomassie blue staining (A) and activity staining (B) following gel electrophoresis. The spot indicated in (B) was excised, digested with trypsin and the resultant tryptic peptides were subjected to LC/ESI-MS/MS analysis, which revealed five peptides (C) that all matched to DLDH after computer database searches (BLAST).

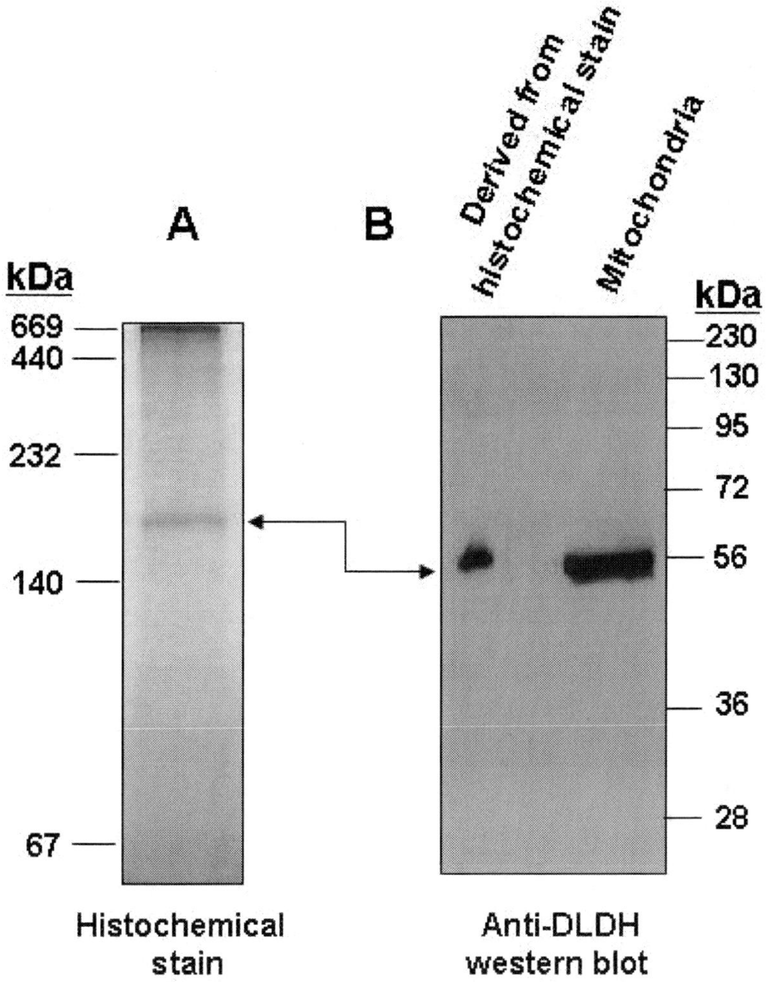

Figure 3. Anti-DLDH Western blot confirming that the diaphorase activity staining band was that of DLDH. Shown are (A) diaphorase activity staining and (B) anti-DLDH Western blot. Lane 1 shows the band derived from BN-PAGE as indicated in (A) that was excised and placed onto a second dimensional SDS-PAGE (10% resolving gel). Lane 2 shows the DLDH immunostaining band derived from mitochondrial extracts.

2.3 Protein amount- and time-dependent DLDH diaphorase activity staining

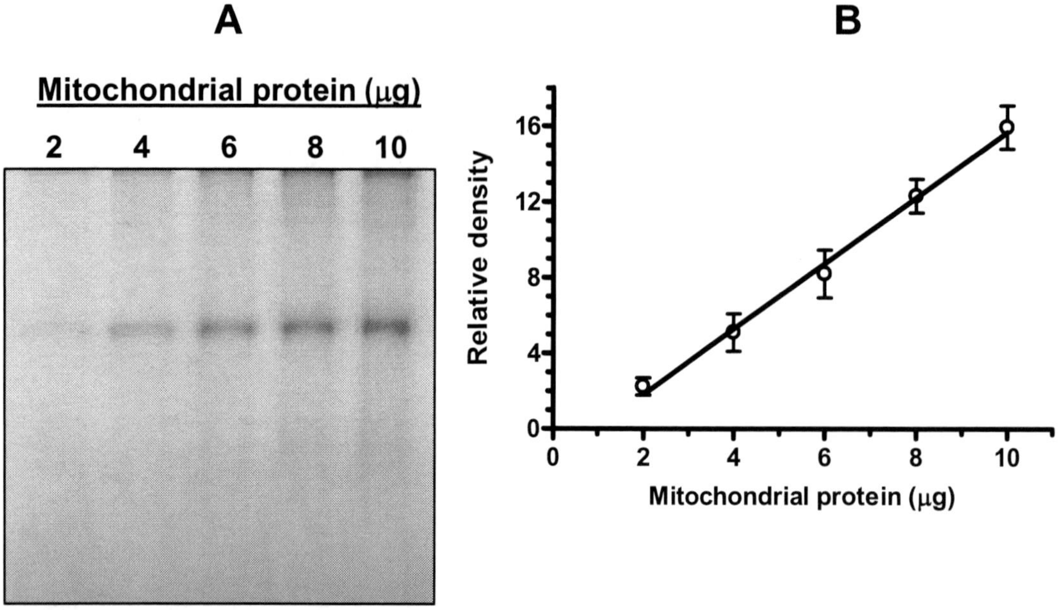

Figure 4. Protein amount-dependent activity staining of DLDH diaphorase activity. Shown are (A) a representative activity staining image and (B) the plot of relative density versus protein amount loaded. Values are mean ± SEM of four independent experiments. Gel was stained by NADH/NBT for10 min.

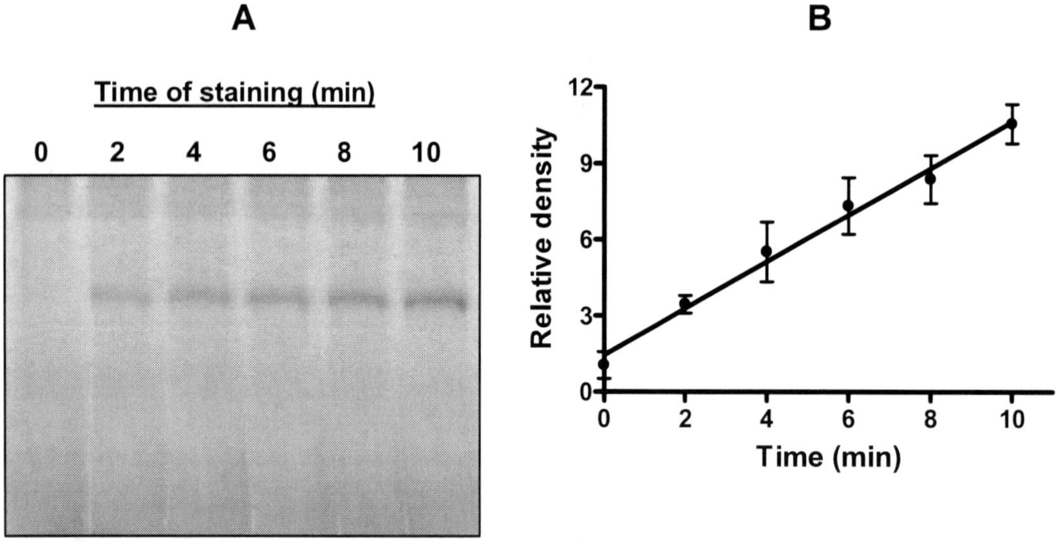

Figure 5. Time-dependent activity staining of DLDH diaphorase activity. Shown are (A) a representative activity staining image and (B) the plot of relative density versus staining time. Following BN-PAGE, each band was excised and stained for the indicated time and all the bands were then placed back together for gel image documentation. Ten microgram mitochondrial proteins were loaded in each lane. Values are mean ± SEM of four independent experiments.

2.4 Comparison of the gel-based method with a spectrophotometric method measuring DLDH dehydrogenas

Figure 6. Comparison of DLDH dehydrogenase activity measured spectrophotometrically and DLDH diaphorase activity determined by the histochemical staining method as described in this paper. The spectrophotometric measurement was performed by monitoring the increase of NADH absorbance at 340 nm upon DLDH-mediated, NAD^+-dependent oxidation of dihydrolipoamide. One unit of dehydrogenase activity is defined as 1 µmol of NAD^+ reduced per min. A solution containing all assay components except mitochondrial sample was used as the blank. Values are mean ± SEM of 3-5 experiments. Upper line: DLDH dehydrogenase activity staining; lower line: BN-PAGE diaphorase activity staining.

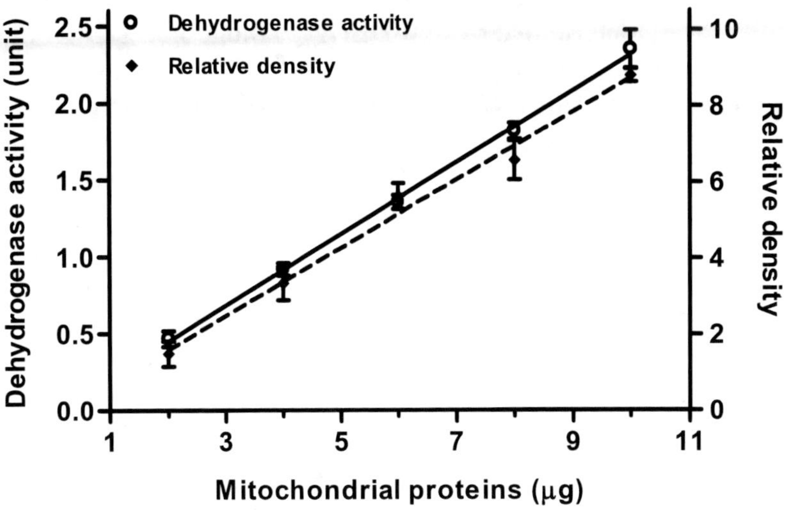

2.5 Applications of the method

2.5.1 Tissue level of DLDH diaphorase activity

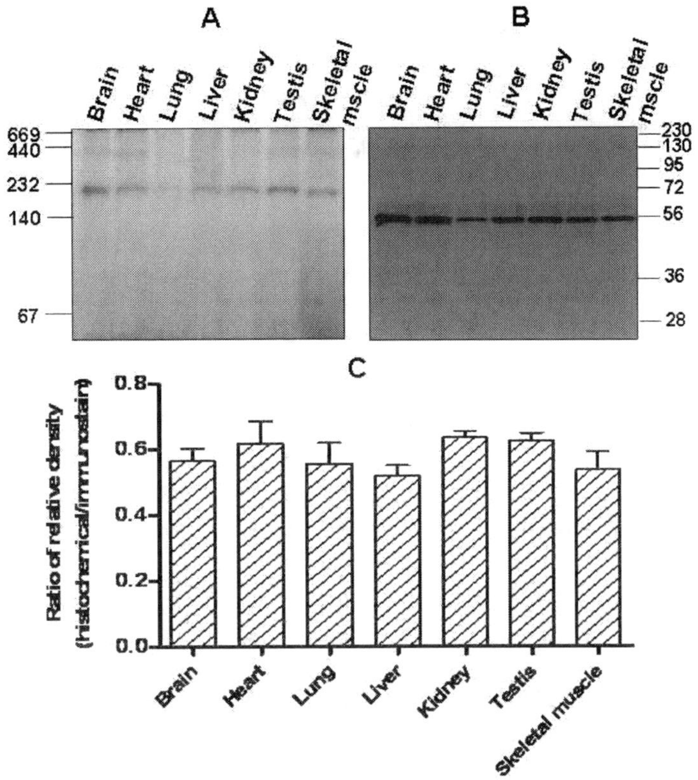

Figure 7. Level of DLDH diaphorase activity in rat tissues. DLDH protein level was determined by anti-DLDH immunoblot. DLDH diaphorase activity was determined by activity staining following BN-PAGE resolution of mitochondrial proteins. Shown are (A) anti-DLDH immunoblot; (B) in-gel activity staining of DLDH diaphorase following BN-PAGE; (D) densitometric quantification of DLDH diaphorase activity expressed as the ratio between activity staining intensity and DLDH immunostain intensity. Values are mean ± SEM of four independent experiments.

2.5.2 Effects of thiol-reactive chemicals on DLDH diaphorase activity

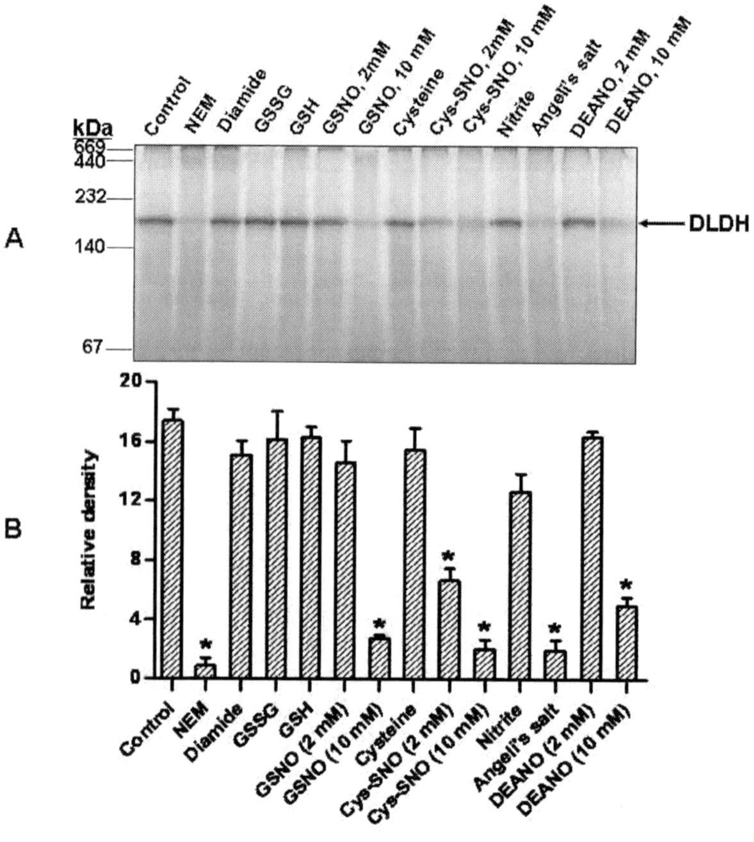

Figure 8. Diaphorase activity staining of DLDH following treatment with the indicated chemicals. Mitochondrial extracts (0.5 μg/μl in a total volume of 20 μl) was incubated with each of the chemicals for 30 min at room temperature followed by immediate mixing with BN-PAGE loading buffer and loading. Where not indicated, a 2 mM concentration (final) was used for the given chemical. Shown are (A) DLDH diaphorase activity staining; (B) densitometric quantification of DLDH diaphorase activity. Control sample (the first lane from the left) was without any treatment. GSH and cysteine were used as additional controls for the respective chemicals. Values are mean ± SEM of four independent experiments. Abbreviations used: GSNO, nitrosoglutathione; CysNO, nitrosocysteine; DEANO, diethylamine NONOate; NEM, N-ethylmaleimide. *$p<0.05$ indicates statistical significance (Welch's t test).

Summary: A BN-PAGE-based method for histochemical staining and quantification of DLDH diaphorase activity is presented. The method involves the resolution of mitochondrial proteins by BN-PAGE that is capable of localizing a DLDH protein band. Diaphorase activity staining is then achieved by DLDH-mediated formation of NBT formazan at the expense of NADH. The technique should be the method of choice when DLDH dehydrogenase activity can not be measured spectrophotometrically due to the presence of interfering chemicals or reactive reagents in assay mixtures. Moreover, the method can also be used for isolation of mitochondrial DLDH that is to be analyzed by other techniques such as mass spectrometry in studying DLDH post-translational modifications.

[1] Williams, C.H.J., in F. Muller (Editor), *Chemistry and Biochemistry of Flavoenzymes*, CRC Press, Boca Raton, 1992, pp. 121-212.

[2] Patel, M.S., Roche, T.E., *FASEB J.* 1990, *4*, 3224-3233.

[3] Neuburger, M., Polidori, A.M., Pietre, E., Faure, M., et al., *Eur J Biochem.* 2000, *267*, 2882-2889.

[4] Faure, M., Bourguignon, J., Neuburger, M., MacHerel, D., et al., *Eur J Biochem.* 2000, *267*, 2890-2898.

[5] Thorpe, C., Williams, C.H., Jr., *J Biol Chem.* 1976, *251*, 3553-3557.

[6] Brautigam, C.A., Chuang, J.L., Tomchick, D.R., Machius, M., et al., *J Mol Biol.* 2005, *350*, 543-552.

[7] Vettakkorumakankav, N.N., Patel, M.S., *Indian J Biochem Biophys.* 1996, *33*, 168-176.

[8] Massey, V., Gibson, Q.H., Veeger, C., *Biochem J.* 1960, *77*, 341-351.

[9] Igamberdiev, A.U., Bykova, N.V., Ens, W., Hill, R.D., *FEBS Lett.* 2004, *568*, 146-150.

[10] Olsson, J.M., Xia, L., Eriksson, L.C., Bjornstedt, M., *FEBS Lett.* 1999, *448*, 190-192.

[11] Xia, L., Bjornstedt, M., Nordman, T., Eriksson, L.C., et al., *Eur J Biochem.* 2001, *268*, 1486-1490.

[12] Patel, M.S., Vettakkorumakankav, N.N., Liu, T.C., *Methods Enzymol.* 1995, *252*, 186-195.

[13] Scouten, W.H., McManus, I.R., *Biochim Biophys Acta.* 1971, *227*, 248-263.

[14] Sokatch, J.R., McCully, V., Gebrosky, J., Sokatch, D.J., *J Bacteriol.* 1981, *148*, 639-646.

[15] Schagger, H., von Jagow, G., *Anal Biochem.* 1991, *199*, 223-231.

[16] Nijtmans, L.G., Henderson, N.S., Holt, I.J., *Methods.* 2002, *26*, 327-334.

[17] Wittig, I., Schagger, H., *Proteomics.* 2005, *5*, 4338-4346.

[18] Singh, R., Chenier, D., Beriault, R., Mailloux, R., et al., *J Biochem Biophys Methods.* 2005, *64*, 189-199.

[19] Beriault, R., Chenier, D., Singh, R., Middaugh, J., et al., *Electrophoresis.* 2005, *26*, 2892-2897.

[20] Novakova, Z., Man, P., Novak, P., Hozak, P., et al., *Electrophoresis.* 2006, *27*, 1277-1287.

[21] Camacho-Carvajal, M.M., Wollscheid, B., Aebersold, R., Steimle, V., et al., *Mol Cell Proteomics.* 2004, *3*, 176-182.

[22] Bautista, J., Corpas, R., Ramos, R., Cremades, O., et al., *Biochem Biophys Res Commun.* 2000, *275*, 890-894.

[23] Jung, C., Higgins, C.M., Xu, Z., *Anal Biochem*. 2000, *286*, 214-223.
[24] Brookes, P.S., Pinner, A., Ramachandran, A., Coward, L., et al., *Proteomics*. 2002, *2*, 969-977.
[25] Sabar, M., Balk, J., Leaver, C.J., *Plant J*. 2005, *44*, 893-901.
[26] Wittig, I., Braun, H.-P., Schagger, H., *Nature Protocols*. 2006, *1*, 418-428.
[27] Swamy, M., Siegers, G.M., Minguet, S., Wollscheid, B., et al., *Sci STKE*. 2006, *345*, pl4.

152g A Strategy for Enriching Lower Abundance Proteins for Two-Dimensional Gel Electrophoretic Analysis

Janice Simler, Bioscience Division, Millipore, 17 Cherry Hill Drive, Danvers, MA 01923

The plasma proteome is a complex sample matrix that contains a wealth of information and serves as a vast source for biomarker discovery. Analysis of the complex plasma proteome by techniques such as two-dimensional (2-D) gel electrophoresis and mass spectrometry is hindered by the presence of highly abundant proteins including albumin and immunoglobulins. Ultrafiltration (UF), with centrifugal devices, has been used to prepare low molecular weight (LMW, < 30kDa) samples for biomarker analysis.

In this report we describe the use of centrifugal UF devices to fractionate both, LMW and HMW, fractions from serum. Protein separation was accomplished by serial filtration through decreasing molecular weight cutoff (MWCO) devices from 100K, 50K, 30K, and 10K MWCO. This serial fractionation strategy enabled the enrichment of proteins by size as well as increasing throughput speed as compared to directly preparing samples using a lower MWCO device. Additionally, the relative purity of samples prepared by the serial fractionation strategy was greater than the relative purity of samples prepared using a single low MWCO device.

Analysis of the LMW proteome by 2-D gel electrophoresis can be improved by using proteins fractionated by size. The serial fractionations method enabled the enrichment of low abundance proteins in the LMW proteome. Further analysis may be done by peptide digestion and mass spectrometry. In conclusion, protein fractionation using centrifugal UF devices is a simple and rapid method to reduce sample complexity and enrich for the low abundance proteins based on molecular weight.

152h Advances In Sample Preparation In The Proteomics Era: Old, Very Old, And Very New Tools For Optimizing Protein Isolation

Gary B. Smejkal, Pressure BioSciences, 6 Gill Street, Suite H, Woburn, MA 01801

The fast, efficient, and accurate release of proteins from cells and tissues is a critically important initial step in most analytical processes, and is essential to reliable proteomic analyses. Two-dimensional gel electrophoresis (2DGE) can be an accurate representation of a proteome only if the entire protein constituency of cells is recovered during the sample preparation process. The use of chaotropes and new surface-reactive agents has substantially improved the capacity of 2DGE to produce more comprehensive maps of cellular and subcellular proteomes, but this increased stringency frequently complicates downstream analyses such as electrophoresis, chromatography, and mass spectrometry. Further, the disruption of cells and tissues usually requires coupling these chemistries with mechanical disruption methods which may contribute to deleterious effect. Pressure Cycling Technology (PCT) uses alternating cycles of high and low hydrostatic pressure to effectively induce the lysis of cells and tissues in preparation for 2DGE and other analytical or preparative methods. Pulses of high hydrostatic pressure perturb equilibria between cellular components causing them to dissociate without breaking covalent bonds. PCT combined with the appropriate extraction buffers is particularly efficient for the dissolution of membrane proteins and other traditionally difficult analytes. PCT derived 37% more protein from the nematode Caenorhabditis elegans than sonication and 17.1% more protein from the gram-negative bacteria Rhodopseudomonas palustris than enzymatic lysis. Remarkably, PCT enabled 2DGE analysis from a single Daphnia pulex organism where over 500 protein spots were detected in silver stained gels. For mammalian tissues, PCT isolated more protein from liver and adipose, both quantitatively and qualitatively, as demonstrated by the appearance of several unique proteins in 2DGE which were not recovered by conventional homogenization techniques. Improved protein recoveries have also been demonstrated for plant tissues where PCT yielded three times more protein from Strelitzia reginae sepals than a centrifugal homogenizer. Moreover, PCT facilitates precise control over sample processing conditions and ensures greater reproducibility.

Web Page: www.pressurebiosciences.com

212a Remotely Powered Microfluidic Pumps And Mixers Based On Miniature Diodes

Suk Tai Chang[1], Orlin D. Velev[1], Erin M. Beaumont[2], Vesselin N. Paunov[3], and **Dimiter N. Petsev**[2]. (1) Department of Chemical and Biomolecular Engineering, North Carolina State University, 911 Partners Way, Centennial Campus, NCSU, Raleigh, NC 27511, (2) Department of Chemical & Nuclear Engineering, UNM, 209 Farris Engineering Center, Albuquerque, NM 87131, (3) University of Hull, Surfactant & Colloid Group, Hull, HU6 7RX, UK

The precise control of fluid transport in microchannels is of paramount importance for the successful design and operation of fluidic devices. We have recently demonstrated1 that using miniature diodes embedded in the walls of fluidic microchannels in combination with AC field is a very simple and convenient tool to manipulate the flows in microchannels. Our focus is on two particular problems briefly described below

Mixing. Due to the low Reynolds number of microflows, mixing of components is a real challenge. Due to the laminar character of the flow different solution tend to flow side by side and the only way for solutes to cross the streamlines is by diffusion. Using properly (anti-parallel) oriented diodes, placed alongside the channel walls, allows generating a vortex fluid motion by simply turning on a properly connected Alternate Current (AC) field source as shown in Figure 1.

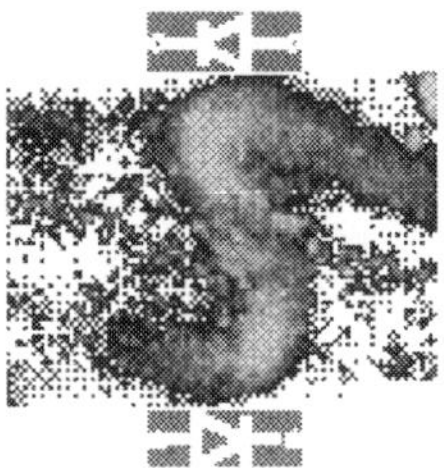

Such vortex dramatically improves the mixing in the microfluidic device. Another advantage of this approach is that such diode mixer can easily be turned on and off through the AC field power source.

Separation. Using parallel oriented diodes and a combination of AC and Direct Current (DC) fields in a loop-shaped channel allows complete decoupling of the fluid electroosmosis from the analyte electrophoresis (Figure 2).

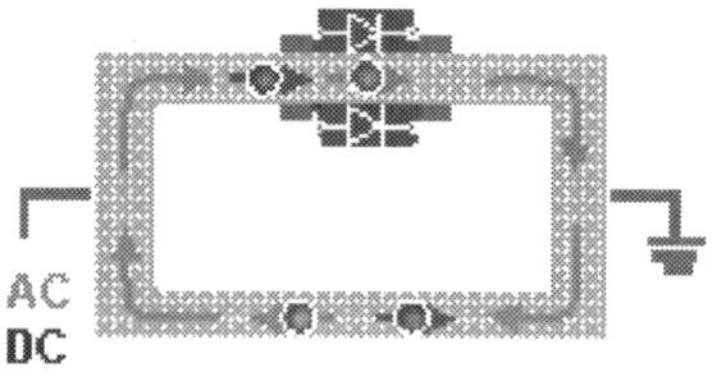

Balancing the electrophoretic and convective forces on the different analytes allows for a very easy and efficient separation in the lower part of the loop in the picture. The parallel oriented pair of diodes, powered by the applied AC field, acts as a miniature pump and drives the fluid in a circulatory motion in the closed loop. Any charged analytes, however, will not migrate in the AC field. Applying DC field to the fluidic device will not drive the fluid motion because this particular design (closed symmetric loop) cancels the electroosmotic driving force. Hence, combining the two fields (AC and DC) allows decoupling of the fluid flow from the particle electrophoretic migration. The electrophoretic migration of the charged analytes will be directed towards one of the electrodes (e.g., left to right – see Figure 2), while the fluid will circulated pumped by the diodes. Therefore, in the lower part of the loop the analyte electrophoresis and fluid flow will be in opposite directions. Independently controlling the AC and DC field magnitudes allows separating species by effectively driving them in opposite directions.

1. S. T. Chang, V. N. Paunov, D. N. Petsev and O. D. Velev, "**Self-Propelling Microdevices and Microfluidic Pumps Based on Remoutely Powered Miniature Semiconductor Diodes**, *Nature Materials*, **6** (2007) p. 235.

Radial Free Flow Electrophoresis for Preparative Separation of Proteins

Jirabhat Lerlertwanich, Prof. Dr.-Ing. Thomas Melin, Dr.-Ing. Süleyman Yüce
Department of Chemical Engineering, RWTH Aachen University, Aachen, Germany

Introduction

Nowadays, electrophoresis is established as a standard laboratory separation method providing high resolution for protein analysis. Nevertheless, its industrial application for preparative protein separation has been so far restricted due to difficulty in scaling up. Different theoretical and experimental studies were devoted to the development of alternative operating modes of electrophoresis in order to overcome such limitations. One promising method is free flow electrophoresis (FFE), which carries out the separation in a flowing liquid film without a matrix medium - as the name "free flow" implies.

Generally, one major scale-up problem of FFE is the necessity to maintain the hydrodynamic stability of the flow. This requires that the temperature gradient caused by Joule heating must be kept sufficiently low so that the forced flow will remain predominant compared to the buoyancy-driven flow [1]. The radial free flow electrophoretic device (Radial FFE) with a narrow separation chamber and intensive cooling at the outer wall can fulfill this requirement.

This work aims at proving the feasibility of the continuous radial FFE device. Preliminary theoretical approaches facilitate the preselection of suitable operational regimes. Experiments using a model protein were carried out to further identify the proper operating conditions regarding electro-hydrodynamic constraints. The performance of the radial FFE in terms of achievable resolution and production capacity was illustrated on an example of a protein mixture.

Materials and Methods

Electrophoretic Apparatus

The radial FFE is described in detail in Fig 1. The parallelepipedic radial separation chamber is 2 mm thick and has an inner and outer diameter of 62 and 312 mm, respectively. Two electrode compartments, located on each side of the chamber, are separated from the separation chamber by 2 mm thick asymmetric ceramic (Al_2O_3) ultrafiltration membranes. The membranes are specifically produced by KERAFOL (Eschenbach, Germany). The top layer of the membranes is made of Mg-Al_2O_3 with an average pore size of 7 nm, while the main body is of 2 µm. At the center of the separation chamber, the outlet assembly, consisting of six concentric slits (50 µm), partitions the total flow into 6 fractions: fraction 1 is the lowest flow layer on the anodic side, represented by "+" in the figure, while fraction 6 collects the uppermost layer near the cathodic side, represented by "-". Each fraction is analyzed individually.

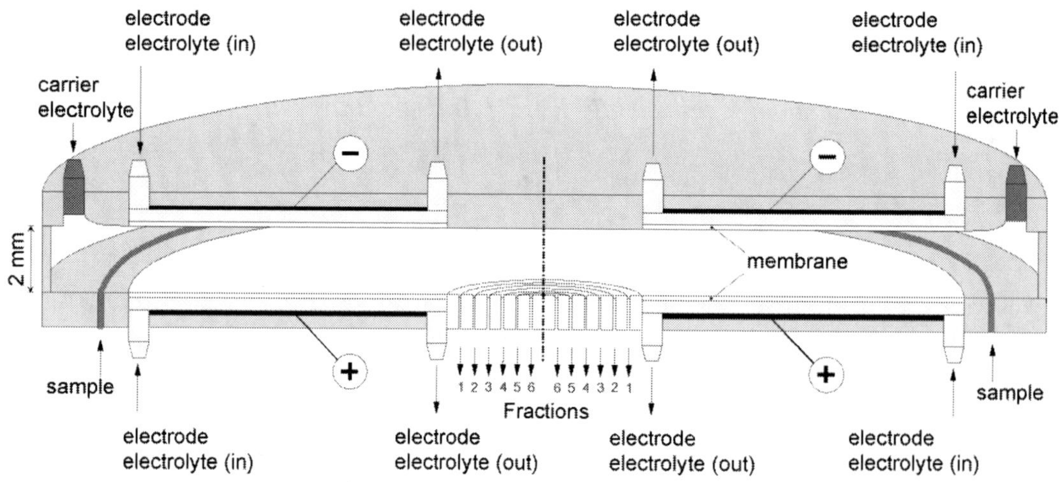

Figure 1. The Radial Free Flow Electrophoretic Device

The carrier buffer is continuously fed over the whole circumference of the separation chamber and flows inwards. The sample is injected from the bottom through a precisely laser-cut slit (70 μm width) into the carrier buffer forming a narrow wall adjacent layer. The electric field is applied perpendicularly to the buffer flow. The sample components are separated laterally in the electric field according to differences in their electrophoretic mobility. The separated components are collected at the outlet.

In order to purge gaseous electrolysis products from the electrode compartments, the electrode electrolytes are continuously circulated at a relatively high flow rate of about 250-300 L/h each and then degassed in a container. Moreover, the electrode electrolytes are used to cool the chamber walls to remove heat caused by Joule heating to avoid free convection within the separation chamber. The operating temperatures are controlled by a LAUDA Ultra-Kryomat thermostat.

Chemicals and Fluids

In all experiments the carrier buffer was a solution of ammonium acetate and acetic acid in de-ionized water with an average pH of 4.6 and an electrical conductivity of 0.5 mS/cm. The electrode electrolytes were also a solution of ammonium acetate and acetic acid (pH 4.6) but at a higher concentration than the carrier buffer, thus having a higher electrical conductivity of 10 mS/cm. The sample solutions were prepared by dissolving pure proteins in the carrier buffer at a concentration of 1 g/L each.

For preliminary experimental investigations, hemoglobin bovine was used as a tracer. The model protein mixture comprised of myoglobin, cytochrome c (from horse heart) and bovine serum albumin (BSA). All substances were purchased from SERVA (Heidelberg, Germany).

Analytical Methods

Electrical conductivity was measured by a digital conductivity meter (Knick Konduktometer 703). pH value was measured with a precision of ± 0.01 with a digital pH-Meter (Knick 647-1). Protein concentration was determined by high pressure liquid chromatography (Agilent), using a GROM-Sil 300 Butyl-1 column with a flow rate of 0.6 mL/min. A digital UV-VIS-spectrophotometer from Varian (Cary 50 Bio) with 6 flow-through cells was used to identify the steady state of the radial

FFE device and for a direct measurement of the hemoglobin concentration. The absorbance of hemoglobin was measured at a wavelength of 406 nm.

Theoretical Approach

Particle trajectory of proteins

In order to estimate a residence time and an electric field strength required for protein separation, a simplified particle trajectory of proteins in a two-dimensional flow filed is calculated. The local velocity of the carrier buffer and therefore of the protein in the radial flow direction (u_r) is defined by the parabolic velocity profile of laminar flow (Eq.1). $\dot{V}$ is the average volumetric flow rate of the carrier buffer and d is the separation chamber height.

$$u_r(r,z) = \frac{6\dot{V}}{2\pi d}\left(\frac{1}{r}\right)\left(\frac{z^2}{d^2} - \frac{z}{d}\right) \qquad (1)$$

The migration of a protein perpendicular to the main flow field depends on its electrophoretic velocity (u_z) as described by equation 2. Electrophoretic mobility (μ) and electric field strength (E) are assumed to be constant. At this estimating step, other types of protein migration, such as molecular diffusion, are not taken into account.

$$u_z = \mu \cdot E \qquad (2)$$

Figure 2 demonstrates the particle trajectories of hemoglobin and BSA in the radial FFE at two different electric field strengths. With an electric field of 100 V/m, both components are gently drawn upwards and remain almost in the same layer. At 2615 V/m, BSA leaves the chamber about in the middle, while hemoglobin, the component having a higher mobility, is reaching the upper wall exactly at the chamber outlet. The electric field strength of 2615 V/m is therefore an upper boundary for the electric field strength allowed to process this sample. Exceeding this maximum value will cause a protein loss into the membrane.

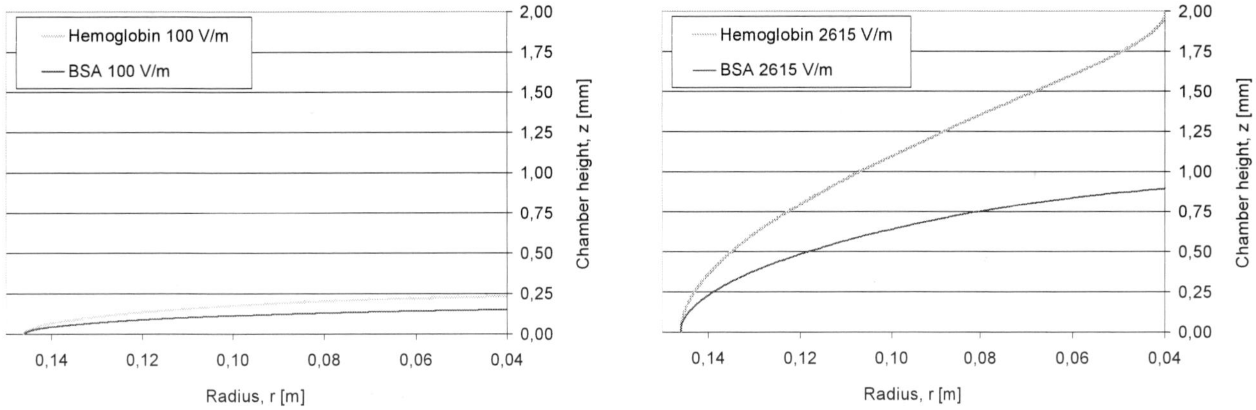

Figure 2: Particle trajectory of hemoglobin and BSA in the radial FFE at 100 V/m and 2615 V/m ($d = 2$ mm, $\dot{V} = 11$ L/h, $\tau = 40$ s, pH = 4.6, $\kappa = 0.5$ mS/cm)

Optimal cooling capacity

The radial FFE device's separation chamber height is only a few millimeters. Heat generated in the carrier buffer is therefore minimal, compared to heat generated in other conventional FFE devices. An adiabatic temperature rise of the carrier buffer operated by the radial FFE is always lower than 1°C

(Eq.3), when assuming no outer wall cooling and not considering further heat dissipation in the membrane.

$$\Delta T = \frac{EiAd}{C_p \rho \dot{V}} \tag{3}$$

However, heat dissipated in the membrane is not negligible. A temperature rise in the membrane can cause a significant temperature profile across the separation chamber and potentially results in thermal convection and thus a severe distortion of the sample stream. In order to define the required cooling capacity, heat removal by electrode electrolytes is modeled and simulated by using COMSOL Multiphysics. The incompressible Navier-Stokes equation (Eq.4) coupled with the conservation law for energy according to equation 5 is used to estimate the velocity and temperature profiles in the separation chamber. The simulation is performed under the flow condition at steady state concerning wall cooling and heat conduction through the membrane (Eq.6).

$$\rho \frac{\partial u}{\partial t} - \nabla \cdot \left[\eta \left(\nabla u + (\nabla u)^T \right) \right] + \rho (u \cdot \nabla) u + \nabla p = F \,, \quad \nabla \cdot u = 0 \tag{4}$$

$$\rho C_p \frac{\partial T}{\partial t} + \nabla \cdot (-k \nabla T) = Q - \rho C_p u \cdot \nabla T \tag{5}$$

$$-\nabla \cdot \left(\kappa \nabla V - J^e \right) = Q_j \tag{6}$$

Further, it is assumed that the electric current density is constant throughout the chamber. A proper cooling capacity in terms of temperature and flow rate of the electrode electrolytes to avoid an onset of hydrodynamic instability is determined. Detailed calculations are presented in the previous work [2].

Results and Discussion

Stable operating regimes

With the theoretical approach presented before, the maximum operational electric field strength can be estimated. In fact, there are other phenomena, which do not allow an operation at that maximum value, because they can cause a sample band broadening or even give rise to hydrodynamic instability of the carrier buffer flow. Such detrimental phenomena are, for instance, molecular diffusion, electro-osmotic flow, and sample distortion due to electro-hydrodynamic effects.

A concentration gradient between the sample and the carrier buffer causes the proteins to diffuse away from the sample layer. The initial narrow sample stream is therefore broadened, especially if the mean residence time in the separation chamber and molecular diffusion coefficient are high. This diffusive broadening is more crucial for small molecules such as amino acids [3].

The electric charge of the membrane depends on a pH-value of the buffer solution filling the membrane pores. If an electric field is applied, the electric non-zero charge double layer on the membrane pore surface results in an electro-osmotic flux through the membrane. A broadening of the sample band occurs if the electro-osmotic flux through the upper and the lower membranes are not equal [3].

Theoretically, the contribution of the electro-hydrodynamic effect to sample distortion is caused by differences in conductivity and dielectric constant between the sample stream and the carrier buffer. For equal dielectric constant, the major axis of distortion is parallel to the electric field if the conductivity of the sample is greater than that of the surrounding carrier buffer, and normal to the field otherwise [4].

In the following, the focus of this work lies on the experimental investigation of the operational constraints due to electro-hydrodynamic distortion, specifically for the radial FFE. Using hemoglobin as a single model protein, experiments were performed with a stepwise increase of the electric current (0A-6A). The electric conductivity of the carrier buffer was 0.5 mS/cm, while the sample with hemoglobin of 1 g/L had a conductivity of 0.53 mS/cm. Figure 3 shows the hemoglobin concentration distribution at different electric currents.

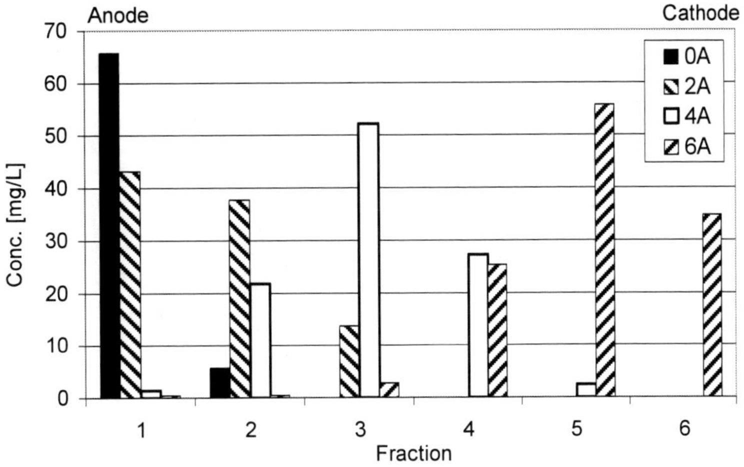

Figure 3. Hemoglobin distribution at different electric currents
(Operating conditions: $\dot{V}_s = 300$ mL/h, $\dot{V} = 11$ L/h, $\tau = 40$ s, $C_0 = 1$g/L)

Without an electric field, haemoglobin ends up only in the first two fractions. With higher electric field strength, the position of the concentration peak moves towards the cathode, while the peak width remains approximately constant (Figure 4). These experimental results show that, up to an electric current of 6 A (E=1940 V/m), no electro-hydrodynamic instability can be observed. The radial FFE can thus be effectively operated at least within this regime.

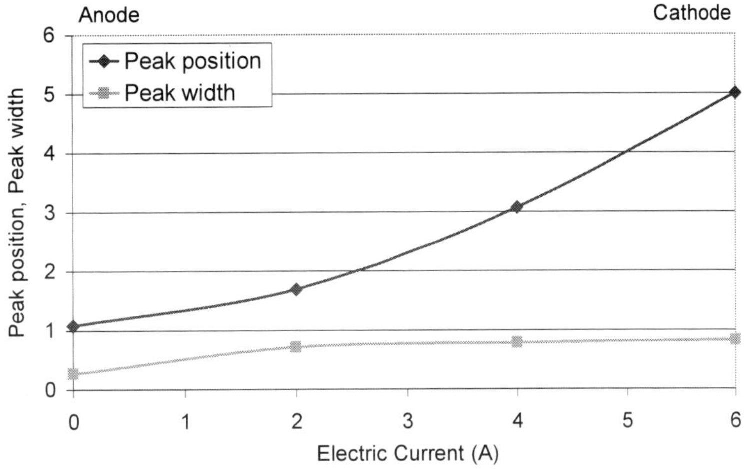

Figure 4. Peak position and peak width of hemoglobin with increasing electric current
(Operating conditions: $\dot{V}_s = 300$ mL/h, $\dot{V} = 11$ L/h, $\tau = 40$ s, $C_0 = 1$g/L)

Separation of a protein mixture

Further experiments were carried out to investigate the achievable resolution and production capacity of the radial FFE by using three model proteins, namely BSA, Cytochrome C and Myoglobin. Analogously to the experiments with hemoglobin, the protein mixture with a throughput of 200 ml/h was separated at various electric field strengths. Figure 5 compares two protein distributions between the case of no electric field (I=0A) and the case of processing with an electric field strength of 1615 V/m (I=5A).

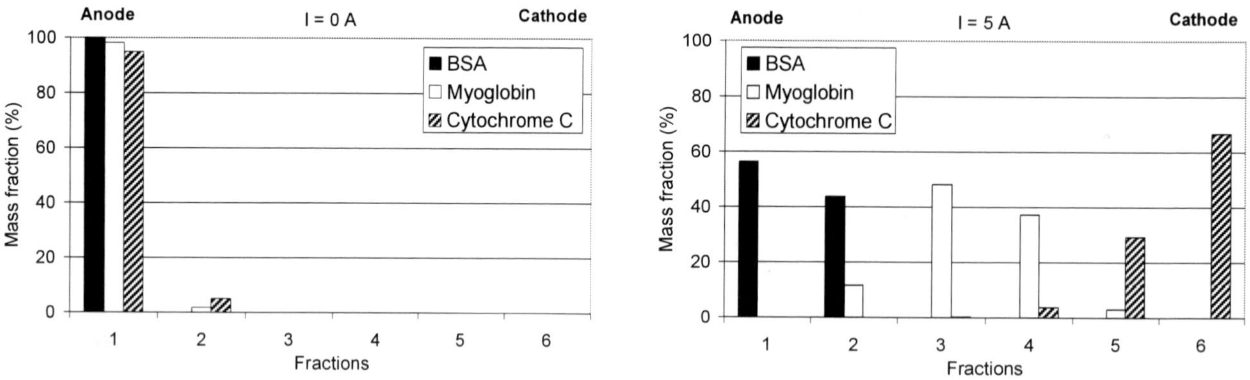

Figure 5. Separation results of a 3-component protein mixture: BSA, Myoglobin, Cytochrome C (Operating conditions: $\dot{V}_s = 200$ ml/h, $\dot{V} = 11$ L/h, $\tau = 40$ s, C_0=1 g/L)

It is shown that, with an electric current of I = 5 A, a relatively clean separation can be achieved. Combining different fractions to product streams yields recovery and purity up to 93% and 97%, respectively (Table 1).

Table 1. Performance of the radial FFE for protein separation

Component	**pI**	**Recovery (%)**	**Purity (%)**
BSA (fraction 1-2)	4.9	85	89
Myoglobin (fraction 3-4)	7.0	93	95
Cytochrome C (fraction 5-6)	10.65	83	97

It is to mention that, this is only a specific performance resulting from processing three components having distinctively different electrophoretic mobilities. For any other separating task, the performance of the radial FFE has to be determined individually. As long as the required electric field strength is kept within the stable operating regimes, the achievable resolution depends mainly on the properties of the proteins in the mixture.

Summary

The radial free flow electrophoretic device is introduced. Due to its narrow separation chamber, a major scale-up problem (Joule heating effect) can be overcome. Theoretical models were proposed to estimate the proper operating conditions. Simplified particle trajectory was calculated to predict the maximum applicable electric field strength, which will not cause protein loss into the membrane wall. In order to avoid hydrodynamic instability caused by Joule heating effect, heat transfer in the device was simulated to determine the temperature and volumetric flow rate of the electrode electrolytes that provide sufficient heat removal. Experiments with a single protein were performed to determine the operating regime, where electro-hydrodynamic instability does not occur. Within this stable regime, a clear separation of a three-components mixture can be achieved.

The authors would like to thank the German Research Foundation (DFG)" for the financial support.

Notation

A	surface area of the electrode	(m^2)	Q_j	electric current source	(Wm^{-3})
E	electric field strength	(Vm^{-1})	t	time	(s)
C	concentration	(gL^{-1})	T	temperature	(K)
Cp	heat capacity	($Jkg^{-1}K^{-1}$)	u	velocity	(ms^{-1})
d	separation chamber thickness	(m)	$\dot{V}$	volumetric flow rate	(m^3s^{-1})
F	force	(N)	$\dot{V}_s$	sample flow rate	(m^3s^{-1})
i	electric current density	(Am^{-2})			
J^e	external generated current density	(Am^{-2})			
k	thermal conductivity	($Wm^{-1}K^{-1}$)	η	viscosity	($kgm^{-1}s^{-1}$)
r, z	coordinates	(m)	κ	electric conductivity	(Sm^{-1})
p	pressure	(Nm^{-2})	μ	electrophoretic mobility	($m^2V^{-1}s^{-1}$)
Q	heat source	(Wm^{-3})	ρ	density	(kgm^3)
			τ	mean residence time	(s)

References

1. Afonso, J.L. and Clifton, M.J. (1998), "Numerical Model of a mixed flow coupled with mass and heat transfer in continuous-flow electrophoresis", Computers & Chemical Engineering, 22 Suppl., pp. S323-S328.
2. Lerlertwanich, J., Melin, T. and Yüce, S. (2006), "Heat transfer analysis to optimize cooling system for a new designed preparative continuous-flow electrophoresis", 17th International Congress of Chemical and Process Engineering (CHISA), Prague, Czech Republic.
3. Poggel, M., Melin, T. (2002), "Transport phenomena in free flow zone electrophoresis", Engineering in Life Sciences., 2(10), pp. 298-304
4. Rhodes, Percy H., Robert, Snyder S. and Roberts, Glyn O. (1989), "Electrohydrodynamic distortion of sample streams in continuous flow electrophoresis", J. of Colloid and Interface Science, 129(1), pp. 78-90

212c Equilibrium Gradient-Focusing Methods

Cornelius F. Ivory, Chemical Engineering, Washington State University, Pullman, WA 99163

Isoelectric focusing (IEF), which is classified by Giddings as an equilibrium gradient method (EGM), is widely considered to be the most powerful technique available for protein separations. Prior to 1985, IEF was the only known electrophoretic EGM but was joined in that year by O'Farrell's Counter-Acting Chromatographic Electrophoresis (CACE). Since that time, more than a dozen new electrophoretic EGMs have been identified, including electric field-gradient focusing and thermal gradient focusing, and it is expected that dozens more may soon join this list.

This paper will begin with a general overview of gradient methods followed by a description of recent advances in this field. I will conclude by suggesting how a new members of the family of orthogonal EGMs might be developed.

212d Electric Field Gradient Focusing In Monolithic Columns Using Nonfouling Materials

Adam T. Woolley, Xuefei Sun, Jikun Liu, and Milton L. Lee. Department of Chemistry and Biochemistry, Brigham Young University, Provo, UT 84602

The need to improve methods for the fractionation of biomolecules in complex mixtures continues to grow. Electric field gradient focusing (EFGF) is a developing technique that allows both separation and enrichment of species such as proteins. EFGF utilizes an electric field gradient combined with laminar flow in the opposite direction to focus molecules into discrete fractions based on their electrophoretic mobilities. We have been evaluating EFGF systems having open capillaries embedded in an ionically conductive polymer, which has a cross-sectional area that varies along the length of the focusing channel [1]. These focusing devices can separate proteins and enrich them by over 10,000-fold. We have also been applying microfabrication methods to construct EFGF systems with smaller channel dimensions, leading to reduced band broadening [2]. Moreover, we have optimized procedures for the fabrication of EFGF systems in poly(ethylene glycol)-based materials, which provide greatly reduced nonspecific adsorption of protein analytes [3]. Finally, we are polymerizing monolithic beds within the focusing column to reduce the band dispersion due to the laminar flow profile. Monolith-filled focusing channels offer enhanced protein separation performance relative to open capillaries of the same dimensions. These improved systems will enable greater capabilities and broader application of EFGF in the analysis of proteins.

References

[1] Humble, P.H; Kelly, R.T.; Woolley, A.T.; Tolley, H.D.; Lee, M.L. *Anal. Chem.* **76**, 5641-5648 (2004).

[2] Kelly, R.T.; Li, Y.; Woolley, A.T. *Anal. Chem.* **78**, 2565-2570 (2006).

[3] Liu, J.; Sun, X.; Lee, M.L. *Anal. Chem.* **79**, 1926-1931 (2007).

Dielectrophoretic Manipulation of Mycobacterium: Toward High Throughput, Post-Mutagenesis Screening for Membrane Lipid Mutations

Benjamin G. Hawkins, Aaron Ezekiel Smith, and Brian J. Kirby, Sibley School of Mechanical & Aerospace Engineering

Introduction

We report the state of development of a dielectrophoresis-based, high-throughput, post mutagenesis screening technique for membrane lipid mutations in *Mycobacterium*. We have successfully actuated M*ycobacterium smegmatis* using continuous-flow, insulative dielectrophoresis; controlling the transverse position of an input stream of cells in glucose media by varying the applied electric field magnitude.

Mycobacterial infectious diseases include leprosy, tuberculosis (TB), and *Mycobacterium avium complex (MAC)*. While incidence in developed nations is low, tubercular infections and fatalities continue to increase worldwide, affecting a significant portion of developing nations[1]. *Mycobacterial* infections kill a continually increasing number of AIDS and other immuno-compromised patients worldwide[1]. In addition to rising rates of infection, the emergence of multi-drug resistant (MDR) strains of both TB and MAC makes the study of these diseases imperative.

Mycobacterial infections are characterized by the formation of encapsulating granulomas – macrophage-encapsulated *Mycobacterium* bacilli surrounded by inflamed tissue – caused by the glycolipids that make up the *Mycobacterium* cell envelope. The mechanisms by which *Mycobacterial* cell wall lipids, and trehelose mycolates in particular, generate this response is not well understood[2]. Determination of these mechanisms requires the development of a library of *Mycobacterium* with mutations in these and other cell wall constituents implicated in the development of *Mycobacteria*'s unique granuloma structure.

While the genome of *M. tuberculosis* is known, the precise function of membrane lipids and their role in the *Mycobacterium*-host cell interactions remains an active area of research[2]. This, as well as the emergence of MDR strains of *Mycobacterium,* illuminates a critical gap in the current toolset used by researchers developing pharmaceuticals and treatment techniques: no library of membrane lipid mutations currently exists for *Mycobacterial* species.

Libraries of mutant strains of *Mycobacteria* and *Escherichia coli* have proven invaluable in the elucidation of virulence factors, biosynthetic pathways, and genomic phenomena[3]. Such libraries consist of samples with single gene mutations. Broad-spectrum screening or focused study of particular mutants facilitates the study of a diverse parameter space; from intracellular molecular processes to host-cell interactions and virulence factors.

The development of such libraries in *Mycobacterium* is frustrated by a lack of direct mutation and mutant screening techniques. *Mycobacterium tuberculosis,* in particular, is difficult to culture *in vitro* and has a long cell cycle time, making development of such a library problematic. Random transposon mutagenesis is one potential technique that can generate a large number of unknown mutations, but no technique exists for identifying those with membrane specific differences in a rapid, high-throughput device, and the time and cost associated with characterizing and sequencing all mutations produced is prohibitive. We present a technique that fills the gap between the generation of phenotyped populations and a library of lipid mutations in *Mycobacterium,* by interrogating numerous cells simultaneously, based on membrane electrical characteristics.

Theory

We predict that *Mycobacterial* envelope properties influence the cell's effective electrical permittivity and conductivity and therefore change a cell's electrical response as a function of membrane composition. We take advantage of this change in electrical character using dielectrophoresis (DEP). Dielectrophoresis – particle motion in response to a non-linear electric field – is induced using insulating constrictions in the depth of a microchannel. The DEP force is proportional to the gradient of the squared electric field, particle volume and the gradient of the squared electric field.

$$\vec{u}_{DEP} = \mu_{DEP} \nabla\left(\vec{E} \cdot \vec{E}\right) \qquad (1)$$

Equation 1 defines the component of particle velocity owing to dielectrophoresis that captures particle properties in a characteristic mobility (μ_{DEP}). Constrictions in channel depth increase the local electric field, causing gradients that induce a local dielectrophoretic force *normal* to the constriction. In addition to non-linear dielectrophoresis, the DC-offset, AC electric field applied in this device (at reservoirs located at microchannel outlets) generates linear electroosmotic fluid flow and linear electrophoretic particle motion everywhere in the device, according to equations 2 and 3:

$$\vec{u}_{EO} = \mu_{EO} \vec{E} \qquad (2)$$

$$\vec{u}_{EP} = \mu_{EP} \vec{E} \qquad (3)$$

In the absence of inertia, we can describe particle velocity as the sum of equations (1), (2), and (3).

Most insulative DEP devices trap particles by countering bulk particle motion (the result of electroomosis, electrophoresis, and/or pressure-driven flow) with local, dielectrophoretic forces, creating a point where the net particle velocity is zero. Insulating constrictions in the width of a microfluidic channel are traditionally used to generate local DEP forces. We, however, constrict the depth of a microchannel, introducing an additional degree of freedom: the angle of incidence between bulk linear electrokinetic forces and opposing nonlinear DEP forces (Figure 1). We then vary the angle of incidence continuously across the channel width, transverse to the direction of flow, by using a

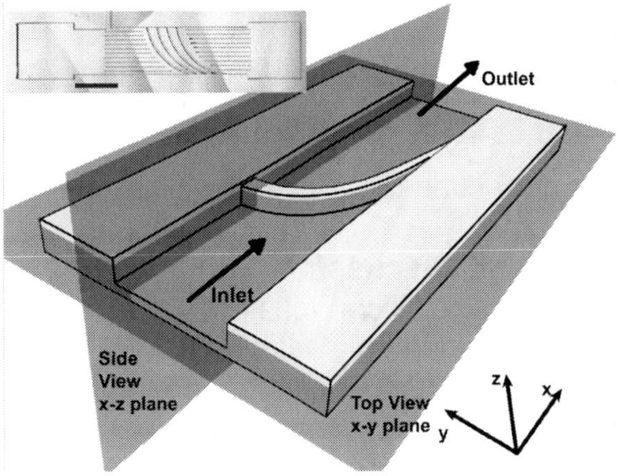

Figure 1: Isometric view of channel and curved constriction. Insulating constrictions in the x-z plane create constrictions in the current path, changing the electric field intensity in these regions and generating the electric field gradients responsible for dielectrophoretic effects perpendicular to the curvature of the constriction. Particle x-axis velocity as a result of electrophoresis, electroosmosis, and dielectrophoresis becomes a function of the angle of the constriction. Inset: A top view of a device embossed in Zeonor. We use horizontal walls at full channel depth to prevent channel collapse and multiple curved ridges to account for any inconsistencies as a result of fabrication error or transient changes in particle properties and local device characteristics.

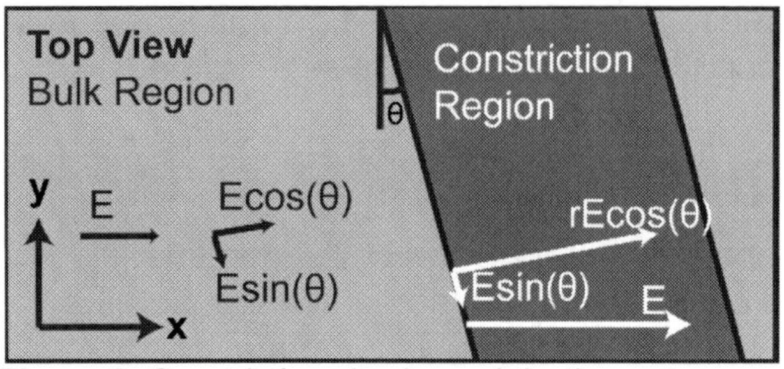

Figure 2: Constrictions in channel depth create electric field gradients that drive DEP forces normal to the constriction. The magnitude of the gradient is therefore dependent on the constriction ratio – bulk channel depth to constriction region depth – and the ridge angle, q. Combined AC and DC electric fields, in the absence of pressure-driven flow, independently modulate linear (electroosmotic and electrophoretic) and non-linear (dielectrophoretic) particle velocity components. We can therefore control linear electrokinetic phenomena by changing the applied DC electric field and non-linear electrokinetic phenomena by controlling the ratio of AC to DC electric fields.

curved constriction in channel depth, creating a DEP velocity component along the x-axis that decreases with transverse position. Particle paths in the presence of this curved constriction are a function of the relative magnitudes of electrophoretic, electroosmotic and dielectrophoretic forces. Particle motion far from the constriction is governed by linear electrokinetic forces. As a particle approaches the constriction, DEP forces are generated perpendicular to the constriction. Depending on particle dependent nonlinear DEP forces, the particle will either pass the ridge or be deflected tangent to the constriction. Particles deflected tangent to the constriction will experience a decreasing DEP force, due to variation in the angle of incidence, and eventually pass over the ridge when linear electrokinetic forces dominate, at a point determined by particle electrical characteristics (Figure 2).

DEP forces on a particle near a constriction are a function of particle properties and constriction angle of incidence. Whether particles pass through or are deflected tangent to the constriction is determined by the ratio of linear electrokinetic forces to non-linear DEP forces, i.e., is a function of particle or cell properties.

Results and Discussion

A two-stage silicon master fabrication procedure followed by hot embossing of microchannels in polymer substrates allows for rapid and facile device fabrication of devices with higher aspect ratios than previously reported. Zeonor polymer substrates are chosen for their optical quality, chemical compatibility with a range of solvents, and durability. After hot embossing the device design from silicon master into Zeonor substrate, the channel is enclosed by a second piece of polymer, and bonded using a chemical bonding technique described previously[5,6]. Confocal microscopy experiments verified the consistency of channel dimensions in constriction regions and the width-to-depth aspect ratio of these devices in the ridge region is greater than 250:1.

Using a numerical solution for the electric field in the vicinity of a curved constriction in channel depth, we can determine the instantaneous particle velocity field using equations (1), (2), and (3). Plotting particle streamlines for a localized input location, we can predict particle trajectories as a function of linear (μ_{EO}, μ_{EP}) and nonlinear (μ_{DEP}) particle mobilities (Figure 3).

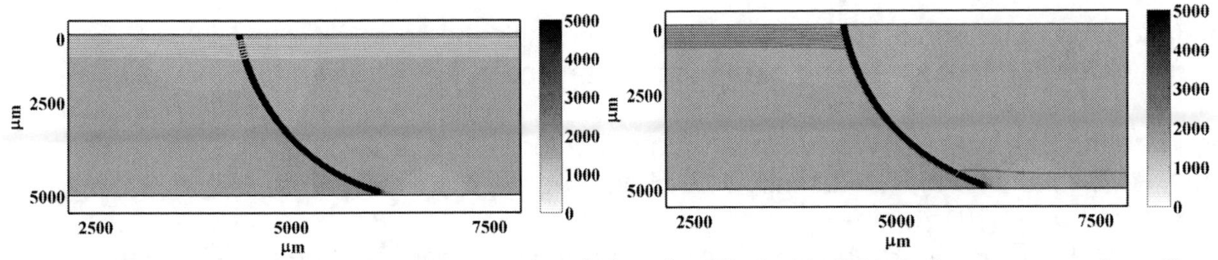

Figure 3: Top view of device, showing numeric solutions for 3D electric fields in the presence of curvilinear constrictions in channel depth, combined with a mobility model for particle motion, allow for rapid, model driven device design for specific applications. The solutions above, for a curvilinear constriction in channel depth, predict particle deflection transverse to the direction of flow as a function of DEP mobility.

We have demonstrated high-throughput separation of polystyrene microparticles with 20% variation in DEP mobility (Figure 4) and continuous control of *M. smegmatis* particle stream position transverse to the direction of flow (Figure 5). *M. smegmatis* is a fast-growing, non-pathogenic bacterium that shares many membrane characteristics with *M. tuberculosis*. In order to characterize particle separation, we examine a time-lapse image of fluorescence intensity and plot the running integral of image intensity at the output (dotted white line in Figure 4). Increasing the AC component of the electric field results in a larger DEP component of particle velocity and increasing deflection of particles transverse to the direction of flow. By comparing these plots, we can clearly see the variation of average particle distributions within the channel as a function of fluorescent wavelength and electric field. Well-characterized test particles, such as polystyrene spheres exhibit predictable behavior, traveling tangential to a curvilinear constriction in depth until linear electrokinetic mobilities dominate and particles pass the constriction at a point transverse to flow characterized by their particular mobility ratio and the operating conditions of the experiment. *M. smegmatis* show the same transverse deflection, increasing with increasing electric field, as observed with micro-spheres. Successes in separation of particles based on DEP mobility, manipulation of *Mycobacterium smegmatis*,

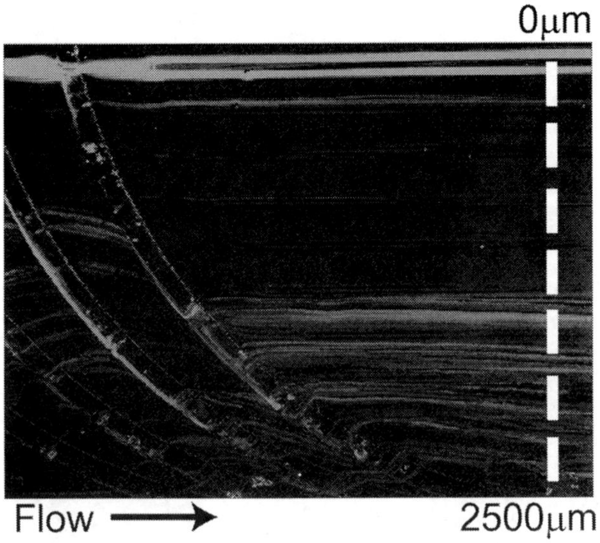

Figure 4: Top view of device showing continuous flow separation of 1.7 and 2.0 micron diameter polystyrene spheres using localized DEP forces to deflect an input stream of particles – image is a time average of 30 frames. Flow direction is from left to right and channel width is 2.5mm.

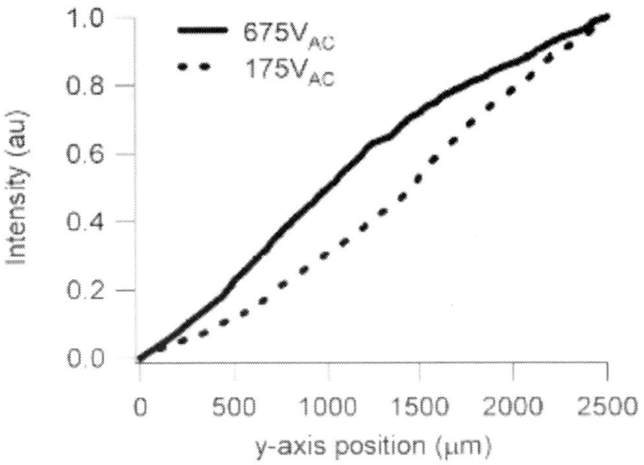

Figure 5: Fluorescently labeled M. smegmatis cells show a variation in output distribution as a function of applied AC electric field. A running integral of image intensity across the channel outlet is used to assess particle distribution as a function of electric field. Increased deflection transverse to flow is a characteristic of increased DEP force applied to particles in the presence of the curved constriction.

and validation through a model-based framework all positively indicate the potential to use this device for high-throughput screening of random mutations in *Mycobacterium*.

[1] World health Organization; World Health Organization, 2006; Vol. 2007.
[2] Geisel, R. E., Sakamoto, K., Russell, D.G., Rhoades, E. R., *The Journal of Immunology*, Apr 2005, 174, 50007-5015.
[3] Rhoades, E. R., R. E. G., Barbara A. Butcher, Sean McDonough, David G. Russell Tuberculosis 2005, 85, 159-176.
[4] Shin, S. J., Wu, C.W., Steinberg, H., Talaat, A.M. *Infection and Immunity*, Jul 2006, 74:7, 3825-3833.
[5] Hawkins, B.G., Smith, A.E., Syed, Y.A., Kirby, B.J. *Analytical Chemistry*, Published online 1 Sept 2007.
[6] Wallow, T. I.; Morales, A. M.; Simmons, B. A.; Hunter, M. C.; Krafcik, K. L.; Domeier, L. A.; Sickafoose, S. M.; Patel, K. D.; Gardea., A. *Lab on a Chip* 2007, Accepted.

212f Electrohydrodynamics In Nanoparticle-Embedded Polymer Gels: Effects Of Morphology And Electrostatic Potential

Jennifer Anne Pascal, Chemical Engineering, Tennessee Tech University, PH-214, Cookeville, TN 38505, Holly A. Stretz, Department of Chemical Engineering, Tennessee Technological University, Prescott Hall, Box 5013, Cookeville, TN 38501, Mario A. Oyanader, Universidad Catolica del Norte, Avenida Angamos, Antofagasta, 0610, Chile, and Pedro E. Arce, Chemical Engineering, Tennessee Technological University, Department of Chemical Engineering, P.O. Box 5013, Cookeville, TN 38505.

Research involving gels with embedded nanoparticles of varying properties is quite attractive because of the multitude of potential applications, including separation of biomacromolecules to aid in the development of new pharmaceuticals, as well as the analysis of the efficiency and control of drug delivery. The presence of nanoparticles within gels has the potential to modify the electrokinetic properties of the gels; therefore, these nanoparticles may influence electro-osmotic flows, as it has been shown in preliminary studies for sensor applications (Matos et al., 2006).

An understanding of the motion of solute within the nanoparticle-embedded gels, such as proteins, DNA, and other macromolecules, can be obtained by performing an analysis that utilizes electrokinetics, hydrodynamics and volume or area averaging (Sauer et al, 1995; Oyanader et. al., 2005). These principles involve the use of electrostatics coupled with convective-diffusive transport. Capillary bundle models are useful for the analysis of the effect of the modified gel architecture on the separation of biomacromolecules in the gels.

This project focuses on capillary models whose characteristics are ideal domains to mimic the gel morphology. Such domains allow for the use of a nonuniform electrostatic potential along the capillary walls in order to capture the electrostatic behavior between the nanoparticles and the gel. They also have the flexibility of incorporating a variable cross section along the axial axis. This communication will report details and illustrations of the modeling efforts. Experimental aspects about the synthesis and characterization of the nano-composite gels are included in a different presentation. (See also, Sedrick, 2007).

References:

Matos, MA, White LR, Tilton RD. Journal of Colloid and Interface Science 2006 (300): 429-436.

Sauer, S., B.R. Locke, and P. Arce, "Effect of Axial and Orthogonal Applied Electric Fields on Solute Transport in Poiseuille Flows: An Area Averaging Approach," Ind. & Eng. Chem. Research, 34, 886 (1995).

Oyanader MA, Arce P. Electrophoresis 2005 (26): 2857.

Sedrick HE, Bollig JR, Stretz HA, Arce P. AIChE Regional Conference, 2007, Columbia, SC.

215a Quick, Easy, Reproducible Protein Isoelectric Fractionation Using Protein Forest *digital* Proteomechip^{TM} Technology

Aaron Sin, James Dasch, William Skea, and Russell Garlick. Protein Forest, Inc., 100 Beaver St, Suite 210, Belmont, MA 02453

Fractionation is necessary to help decipher protein content in a complex sample. Protein Forest has developed a quick, reproducible and easy to use isoelectric fractionation technology: the digital ProteomeChip^{TM} (*d*PC^{TM}).

The *d*PC^{TM} consists of discrete pH gel features with pH intervals of 0.1 or smaller, which, when placed in an electric field for 30 min, will each trap and collect proteins simultaneously from a complex mixture according to the proteins' respective isoelectric points (pI). The resultant protein fractions can then be enzymatically-digested for subsequent MS identification, blotted to a membrane for antibody staining or transferred to traditional SDS-PAGE. The individual pH features allow researchers to reproducibly extract protein fractions from a complex mixture for detailed analysis, such as in biomarker discovery, protein-protein interaction research and clinical assay applications.

We have applied the *d*PC^{TM} technology to fractionate complex cell lysates and analyzed using LC/MS. By combining the *d*PC^{TM} technology with western analysis, we were able to perform pI separation on rat liver cell lysates and quickly determine the post-translational modification states of potential biomarker proteins. In this talk, we will discuss representative experimental results and other potential applications of the *d*PC^{TM} technology.

Web Page: www.proteinforest.com

215b An Orthogonal Affinity Capture and Detection Strategy for Identifying and Characterizing Phosphoproteins Directly from Sds-Polyacrylamide Gels: Inorganic Chemistry to the Rescue!

Alvydas Mikulskis, Linan Song, Alla Bogdanova, Hao Xie, Eva Golenko, Yang Wang, and Wayne F. Patton. Biochemistry, PerkinElmer Life & Analytical Sciences, 940 Winter Street, Waltham, MA 02451

An antibody-free process workflow is described involving orthogonal phosphomonoester-selective binding strategies. First, complex protein samples are enriched for phosphoproteins using an alkoxide-bridged dinuclear zinc complex conjugated to agarose beads under nondenaturing conditions, at neutral pH. The enriched phosphoproteins are separated by conventional SDS-polyacrylamide gel electrophoresis and a fluorophore-conjugated alkoxide-bridged dinuclear zinc complex is employed to selectively highlight phosphoproteins via binding to the phosphomonoester dianion moieties of serine, threonine, and tyrosine. Interaction with other anionic residues, including carboxylate residues on proteins, is insignificant. As little as 1 ng of phosphoprotein is detectable by this method using standard charge-coupled device camera- or laser scanner-based imaging systems. Then, phosphoprotein bands are excised from the gel, subjected to proteolytic digestion and constituent phosphopeptides subsequently purified using Titania thin-film coated magnetic beads at acidic pH. Phosphopeptides are eluted at alkaline pH and directly characterized by MALDI-TOF or tandem mass spectrometry. Phosphopeptides can readily be identified with minimal contamination of samples by acidic peptides, as is often encountered using conventional trivalent ferric- or gallium-based immobilized metal ion affinity chromatography (IMAC) approaches. The workflow does not require two-dimensional gel electrophoresis or nano-liquid chromatography, making analysis of phosphoproteins accessible to a broad spectrum of Life Sciences investigators.

215c Instrumentation and Protocols for Optimized Separation of Peptides by Pressurized Planar Electrochromatography, with Subsequent Direct Interrogation of Analytes by Tandem Mass Spectrometry

Wayne F. Patton[1], David Bourdon[2], Peter Jackson[3], Mei Wang[1], Yang Wang[1], Linan Song[1], Bob Krug[2], and Steven Liebold[2]. (1) Biochemistry, PerkinElmer Life & Analytical Sciences, 940 Winter Street, Waltham, MA 02451, (2) R&D, PerkinElmer Life &Analytical Sciences, 2200 Warrenville Road, Downers Grove, IL 60515, (3) R&D, PerkinElmer Life & Analytical Sciences, Chalfont Road, Seer Green, Beaconsfield, Bucks, United Kingdom

Thin-layer chromatography (TLC) of peptides has not been widely implemented, mainly because alternative techniques are readily available for peptide analysis, such as SDS-polyacrylamide gel electrophoresis, capillary electrophoresis and peptide mass fingerprinting by matrix-assisted laser desorption time-of-flight mass spectrometry (MALDI-TOF MS). Pressurized planar electrochromatography (PPEC) offers much higher separation efficiency and more rapid analyte fractionation in comparison to conventional TLC. Peptide separations by PPEC result from a combination of electroosmotic flow of the mobile phase and electrophoresis of the zwitterionic analytes. A newly developed instrument is presented that enables 85 mm square planar chromatography plates to be uniformly pressurized to 20-30 psi and run by high-voltage PPEC. This plate format is directly compatible with MALDI-TOF MS and MS/MS, conforming to most existing target plate standards. Typically, conventional TLC of peptides has only provided information pertaining to the retardation factor (Rf) of a peptide, which is not sufficient for the unambiguous assignment of sequence information. However, MALDI-TOF MS and MS/MS analysis of peptides directly from the silica surface of the TLC plate after PPEC separation can be performed with high m/z accuracy, enabling peptide sequencing and identification, as well as localization of phosphorylation sites.

215d A Hybrid Methodology For Peptide Identification Via Mixed-Integer Linear Optimization, Local Alignment Database Search And Tandem Mass Spectrometry

Peter A. DiMaggio, Chemical Engineering, Princeton University, Princeton, NJ 08544 and Christodoulos A. Floudas, Department of Chemical Engineering, Princeton University, Engineering Quadrangle, Princeton, NJ 08544.

Peptide and protein identification is of fundamental importance in the study of proteomics. Tandem mass spectrometry (MS/MS) coupled with high performance liquid chromatography (HPLC) has emerged as a powerful protocol for high-throughput and high sensitivity peptide and protein identification experiments. In recognition of the extensive amount of sequence information embedded in a single mass spectrum, tandem MS has served as an impetus for the recent development of numerous computational approaches formulated to sequence peptides robustly and efficiently with particular emphasis on the integration of these algorithms into a high throughput computational framework for proteomics. The two most frequent computational approaches reported in literature are (a) de novo and (b) database search methods, both of which can utilize deterministic, probabilistic and/or stochastic solution techniques. De novo methods have distinct advantages over database methods in that they can analyze peptides not present in a protein database and are more amenable to identifying post-translational modifications. Recent work has been done to combine the strengths of de novo and database techniques to improve peptide identification accuracy [1-4]. A typical framework utilizes the de novo sequences to generate large "sequence tags" which are used to query a protein database. Such a framework has the advantage of verifying statistically insignificant database identifications and/or resolving residue ambiguities in the de novo predictions.

We have recently developed a novel mixed-integer linear optimization (MILP) approach to efficiently address the de novo peptide identification problem so as to form a basis for a high-throughput computational framework for peptide identification [6, 7]. This framework is denoted as PILOT, which stands for Peptide identification via Integer Linear Optimization and Tandem mass spectrometry. The overall algorithm PILOT is comprised of: (1) a preprocessing algorithm used to identify certain peaks and to validate boundary conditions, (2) a two-stage mixed-integer linear optimization framework to address missing ion peaks due to residue-dependent fragmentation characteristics, and (3) a post-processing technique for selecting the most probable sequence by cross-correlating the theoretical spectra of the candidate sequences with the experimental tandem mass spectrum.

In this work, we propose a hybrid methodology which utilizes the rank-ordered list of de novo predictions provided by PILOT to query the non-redundant protein database using FASTA [7]. High-confidence residues from the de novo predictions are identified based on peak intensities, existence of complementary ions and conservation of subsequences over all possible candidate sequences. A modified BLOSUM scoring matrix is constructed in order to bias exact residue matches in the alignment with an additional award for high-confidence residues in the query sequences. The computational burden associated with performing several sequence alignment calculations is circumvented by use of parallel programming. The individual run time for each sequence alignments is also reduced by reformatting the non-redundant protein database. Results for the hybrid methodology for peptide identification will be presented on experimentally validated quadrupole time-of-flight (QTOF) tandem MS [8], a test set of annotated OrbiTrap tandem MS, and a bench mark set of ion trap tandem MS [9].

[1] J. Taylor and R. Johnson. Sequence database searches via de novo peptide sequencing by tandem mass spectrometry. Rapid Commun. Mass Spectrom., 11: 1067-1075, 1997.

[2] A. Shevchenko, S. Sunyaev, A. Loboda, A. Shevchenko, P. Bork, W. Ens, and K. Standing. Charting the proteomes of organisms with unsequenced genomes by MALDI-quadrupole time-of-flight mass spectrometry and BLAST homology searching. Anal. Chem., 73: 1917-1926, 2001.

[3] N. Wielsch, H. Thomas, V. Surendranath, P. Waridel, A. Frank, P. Pevzner, and A. Shevchenko. Rapid validation of protein identifications with borderline statistical confidence via de novo sequencing and MS BLAST searches. J. Proteome Res., 5: 2448-2456, 2006.

[4] A. Frank, M. Savitski, M. Nielsen, R. Zubarev, and P. Pevzner. De novo peptide sequencing and identification with precision mass spectrometry. J. Proteome Res., 6: 114-123, 2007.

[5] P.A. DiMaggio and C.A. Floudas. A mixed-integer optimization framework for de novo peptide identification. AIChE Journal, 53(1), 160-173 (2007).

[6] P.A. DiMaggio and C.A. Floudas. De novo peptide identification via tandem mass spectrometry and integer linear optimization. Anal. Chem., 79, 1433-1446 (2007).

[7] W. Pearson and D. Lipman. Improved tools for biological sequence comparison. PNAS, 85:2444-2448, 1988.

[8] B. Ma, , K.Z. Zhang, C. Hendrie, C. Liang, M. Li, A. Doherty-Kirby, and G. Lajoie. PEAKS: powerful software for peptide de novo sequencing by tandem mass spectrometry. Rapid Commun. Mass Spectrom., 17:2337-2342, 2003.

[9] A. Frank and P. Pevzner. Pepnovo: De novo peptide sequencing via probabilistic network modeling. Anal. Chem., 77(4):964-973, 2005.

215e Integrated Top-Down And Bottom-Up Strategy For Characterizing Intact Proteins And Their Modifications

Ljiljana Pasa-Tolic, Biological Sciences Division, Pacific Northwest National Laboratory, 902 Battelle Boulevard, P.O. Box 999 M.S. K8-98, Richland, WA 99352

Two complementary strategies for protein characterization are typically employed using mass spectrometry: bottom-up and top-down, each with its own strengths and weaknesses. We present an integrated top-down / bottom-up approach facilitated by concurrent LC/MS analysis and fraction collection. For comprehensive high-throughput intact protein profiling, we employed reversed phase LC separations coupled on-line with a 12T Fourier transform ion cyclotron resonance (FTICR) mass spectrometry. Confident assignments of intact protein masses and their gas phase fragment spectra are enabled by limiting the bottom-up analysis to individual fractions and thereby reducing the complexity of the protein mixture undergoing analysis. For large-scale proteome profiling, additional dimension of protein separation technique, such as free flow electrophoresis (FFE), might be employed to reduce the complexity and improve the quality of MS. Herein, we apply this novel comprehensive strategy to elucidate various biological phenomena such as dynamics of protein modifications under conditions of oxidative stress.

215f ^{15}N Metabolic Labeling And Tandem Mass Spectrometry: Tools For The Proteome Characterization Of The Facultative Psychrophile *pedobacter Cryoconitis* Via Ortholog Searching And *de Novo* Sequencing

Ana G. Pereira-Medrano[1], Rosa Margesin[2], and Phillip C. Wright[1]. (1) Department of Chemical and Process Engineering, The University of Sheffield, Biological & Environmental Systems Group, Sheffield, S1 3JD, United Kingdom, (2) Institut für Mikrobiologie, Leopold-Franzens-Universität, Technikerstraße 25, Innsbruck, A-6020, Austria

Proteomics is becoming increasingly important in analyzing biological samples from the environment. This area is still relatively embryonic due to challenges associated with sample preparation difficulties and the absence of a genome sequence. However, new informatics tools and mass spectrometry techniques are allowing for strides to be made in this area. Here, we present the use of ^{15}N metabolic labeling and tandem mass spectrometry for the proteomic analysis of a facultative psychrophile from alpine glacier cryoconite, *Pedobacter cryoconitis*[1]. This strain has the ability to cometabolically degrade petroleum hydrocarbons, which had not yet been described for other *Pedobacter* species. The genomic sequence of this strain is not available. A method for protein identification and quantitation using stable isotopes via 15N metabolic labeling was carried out. The impacts of increasing temperature and different carbon sources were investigated.

Replicate cultures of *P. cryoconitis* were grown at two different temperatures (1 and 20°C) and with two different carbon sources (maltose and glucose) in mineral medium with ^{15}N-labeled ammonium sulphate (20°C) and unlabeled ammonium sulphate (1°C). Cells were harvested at the mid-late log growth phase, proteins extracted from centrifuged pellets with liquid nitrogen cracking, and stored at -20oC. Each sample and biological replicate was then treated as follows: 2DGE (1000 µg in 3-10 pH IPG strip), gel analysis using Progenesis SameSpots v2.0 to determine most intense spots present in all replicates, followed by in-gel tryptic digestion, LC-MS/MS analysis with an Applied Biosystems QStarXL, database searches with a Nitrogen constraint using Mascot, and results compared to *de novo* sequencing and protein identification with PEAKS and MS Blast.

On average, 98 spots were each excised from 4 gels (2 biological replicates of cells grown at 1°C and at 20°C with glucose or maltose as the sole carbon source) resulting in 392 spots. One protein per spot was identified by searching for peptide orthologs from homologous proteins with Mascot and PEAKS Studio 4.2 protein identification (NCBI nr bacteria database of sequenced organisms) in both replicates. The peptides in these proteins were then given extra validation by searching the TOF-MS spectrum for the ^{15}N doublet peak being the requisite m/z distance away based on the number of nitrogen atoms present in the unlabelled peptide. The N constraint aids in the unambiguous identification and relative quantification of proteins by determining the nitrogen composition of both intact peptides and fragments, without the need for precise mass spectrometry. Identified proteins using the N constraint showed typical MOWSE scores around 80 or higher based on two or more identified peptides. PEAKS

Studio 4.2 was used to perform *de novo* sequencing on spectra of spots with no validated protein identification by ortholog searching. *De novo peptide* sequences were then searched with NCBI Protein Blast as a sequence similarity searching protocol to obtain a protein similarity results. MS Blast scores were typically 60 or higher.

The relative quantitation of these peptides was performed by examining the relative areas under the doublet peaks of each peptide in the TOF-MS spectrum. Protein identification and relative quantitation was achieved. Example proteins found to be differentially expressed include 60 kDa chaperonin and translation elongation factor EF-Tu, both upregulated at 20°C. This data provides the first example of a proteome characterization of this unsequenced psychrophile employing 2 dimensional gel electrophoresis (2DGE), tandem mass spectrometry, ^{15}N metabolic labeling, nitrogen-constrained ortholog searching, and *de novo* sequencing.

[1] Margesin, R., Sproer, C., Schumann, P. and Schinner, F., Pedobacter cryoconitis sp. nov., a facultative psychrophile from alpine glacier cryoconite. Int J Syst Evol Microbiol. 2003, 53, 1291-1296.

215g Identification Of Phosphotyrosine-Containing Proteins By Immunoprecipitation And Mass Spectrometry

Nancy Kendrick, Jon J Johansen, and Matthew Hoelter. Kendrick Labs Inc, 1202 Ann St, Madison, WI 53713

The PY20 antibody against phosphotyrosine is useful for studying the mechanism of action of phosphotyrosine kinases using the method of 2D gel Western blotting (WB). Differences between control and treated samples may easily be found on 2D gels because the WB patterns are much simpler than the very complex patterns obtained from conventional staining. If the PVDF blot is stained with Coomassie blue and scanned prior to immunostaining, matching between the ECL film and the Coomassie pattern of a duplicate 2D gel becomes straightforward. Proteins of interest may then be cut from the duplicate Coomassie gel and identified by mass spectrometry (MS).

However, in many cases these important proteins are of too low an abundance to be identified from the original sample. In this paper we describe a method for immunoprecipitation of phosphotryosine-containing proteins from large amounts of sample prepared for 2D electrophoresis, in order to get enough material for MS. Approaches, problems and results will be presented.

Web Page: www.kendricklabs.com

215h Fabrication Of Cell-Free Expressed Protein Libraries For Proteomics: The Use Of T-RNA Engineering, Photocleavable Linkers And Multiplexed Protein Production On Beads

Mark J. Lim, AmberGen Inc., 100 Beaver Street, Room 269, Waltham, MA 02453, Sadanand Gite, Molecular Diagnostics, AmberGen Inc., 100 Beaver Street, Waltham, MA 02453, Ziying Liu, Proteomics, AmberGen Inc., 100 Beaver Street, Waltham, MA 02453, and Kenneth J. Rothschild, Department of Physics and Photonics Center - Molecular Biophysics Laboratory, Boston University, Boston, MA 02215.

Proteome microarrays hold great promise for a variety of biotechnological applications including mapping protein-protein interactions, measuring drug-protein interactions and biomarker discovery. However, the need to express, purify and print thousands of human proteins at high density on a microarray substrate presents challenges which limit their wide-spread availability and use. We have developed a suite of novel technologies aimed at the production of in vitro (cell-free) expressed protein libraries, both on microarray surfaces or displayed on beads. Expressed proteins are isolated on beads using photocleavable linkers incorporated directly into the nascent protein during its synthesis using tRNA Mediated Protein Engineering (TRAMPE). In a different approach, proteins are isolated using antibodies immobilized on the beads through a photocleavable linker. Proteins can then be photo-released into solution (PC-SNAG) for conventional microarray printing. Alternatively, nascent proteins and coding agents are transferred directly from the beads to a surface to form discrete microarray spots whose dimensions match that of the beads, using a process termed photocleavable printing (PC-PRINT). PC-PRINT can also be used in conjunction with MALDI-MS by transferring proteins to MALDI substrates for downstream mass spectrometry based analyses. TRAMPE incorporated fluorescence labels also allow quantification and tracking of the proteins during their purification and deposition on the solid surface. When combined with methods to create Bead-Sorted Libraries of In Vitro Expressed Proteins (BS-LIVE-PRO), based on multiplexed solid-phase PCR and multiplexed cell-free expression reactions, this new approach provides an inexpensive method to fabricate very large scale, high density proteome microarrays (or bead displays). Furthermore, this multiplexed approach, in conjunction with rapid cell-free protein expression can facilitate the fabrication of readily customizable, on-demand microarrays of selected content. The resultant microarrays are shown to be compatible with measurements of protein-protein interaction. This work was partially funded by the SBIR grant program of the NIH.

273a Temperature-Mediated Purification Of Genomic Material Via Electrophoretic Adsorption Of Proteins Onto Hydrophobic Moieties In A Polymer Matrix

Ryan E. Forster, Materials Science and Engineering, Northwestern University, 2145 Sheridan Road, Technological Institute, E136, Evanston, IL 60208 and Annelise E. Barron, Chemical and Biological Engineering, Northwestern University, 2145 Sheridan Road, Evanston, IL 60208.

On-chip purification of genomic material from crude biological samples such as whole blood is crucial for developing point-of-care field devices for use at home and abroad. Our group has synthesized and previously reported on a series of water-soluble block copolymers containing acrylamide and hydrophobic N,N,-dialkylacrylamides, which can adsorb molecules containing hydrophobic moieties such as proteins during electrophoresis while allowing genomic material to pass through. We have theorized that protein removal is not only dependent on the polymer structure, but also on the environments in which they are used, most notably the running temperature of the system. As hydrophobic interactions generally become stronger with temperature, based on increased exposure of hydrophobic groups due to denaturing and thermodynamic effects, the effectiveness of the polymer networks is shown to be dramatically enhanced with increased temperature. Tests were performed in 10 ° C increments from 25 °C to 65 °C on an 8-cm long glass microfluidic chip dynamically coated with poly (N-hydroxyethyl-acrylamide) (pHEA); the proteins studied were fluorescently labeled with Alexa Fluor 488 and detected using a homebuilt LIF system. Results have shown that by increasing the temperature to 65°C, and utilizing a 600-kDA (Mw) copolymer containing only ~ 0.20 mol% dihexylacrylamide, we can remove a larger percentage of bovine serum albumin proteins than previously possible. Remarkably, proteins that have few exposed hydrophobic amino acids at room temperature, such as trypsin inhibitor, which previously passed through the hydrophobic copolymer network unhindered, has also been successfully removed while allowing DNA to pass through freely. The effectiveness of this technique at elevated temperatures also allows for copolymers containing smaller concentrations of hydrophobe to be utilized while obtaining similar results to those containing a larger percentage. This is crucial, as the viscosity of the polymer solution is sensitive to the percentage of hydrophobic moiety present in the copolymer, and the ability to utilize less hydrophobe at elevated temperatures will decrease pressure and strength requirements for loading the polymer solution into the microchannels at ambient temperature. Based on these results, future testing of crude biological samples as well as the development of microchips and copolymers specifically tailored to this system will be pursued.

273b Band Broadening During High-Throughput Mutation Detection In Microchannels

Nabil Laachi and **Kevin D. Dorfman**. Chemical Engineering and Materials Science, University of Minnesota, 151 Amundson Hall, 421 Washington Ave SE, Minneapolis, MN 55455

Microfluidics and lab-on-a-chip technologies promise great advances in bioanalytical chemistry, most notably for DNA analysis. While most DNA analysis involves electrophoretic sizing, it is also possible to use electrophoresis to detect mutations in identical size DNA based upon the different interactions between wild type (or mutant DNA) and a linear polymer sieving matrix. In denaturant electrophoresis, the presence of a mutation destabilizes the DNA double helix, leading to a lower melting temperature. In the ideal case, the system is operated at a temperature such that the wild type strand is fully annealed, while the mutant strand exhibits a denaturation bubble. The denaturation bubble lowers the electrophoretic mobility of the mutant relative to the wild type, permitting a simple screen for the presence of a potentially harmful mutation. While denaturant electrophoresis techniques are well developed for macroscopic gel-based separations, there are significant challenges associated with downsizing the method. In particular, temperature control becomes a major issue, since the low thermal mass of the microchannel systems and capillaries makes it difficult to ensure a homogeneous temperature, especially for parallel, high-throughput operations. One possible method for circumventing the temperature control problem is to cycle the temperature of the microchannel between a low temperature, at which both the wild type and mutant strands are annealed, and a higher temperature, at which both strands are partially denatured. This method, known as cycling temperature electrophoresis, ensures that every channel operates at some time in the ideal temperature range for detection. We consider theoretically the effect of the temperature oscillations on the efficiency of the separation. Using a multiple-time scales analysis, we compute the mean velocity and dispersivity of the DNA as a function of the mobility in the annealed and denatured states and the kinetics for the denaturation/annealing reactions. Explicit results are presented for a step change in the reaction kinetics. We find that the relaxation to the annealed state can lead to significant band-broadening, and our analysis indicates that the separations will be most efficient at weak electric fields.

Computer-assisted analysis of two-dimensional gel patterns derived from agarose electrophoresis of conjugated Hib meningitis vaccines

Dietmar Tietz

Correspondence: Dr. Dietmar Tietz, 8167 Shoal Creek Drive, Laurel, MD 20724
Email: djt@his.com

Introduction

Although analytical methods for vaccine characterization based on 2-D agarose electrophoresis were developed in the period from 1983 to 1995, they remain a promising tool for vaccine quality control and for predicting vaccine effectiveness in the production of semi-synthetic immunogens (See reference [1] for a recent review).

Vaccine samples and size standards

The protein-polysaccharide conjugated meningitis vaccines were developed by Robbins, Schneerson, and coworkers [2, 3] for the immunization of small children, the main target group of bacterial meningitis caused by *Haemophilus influenzae* type b (Hib). In 1996 Robbins and Schneerson shared a Lasker Clinical Research Award for their groundbreaking work [4].

Earlier vaccine preparations varied in immunogenicity, and immunological tests for identifying potentially effective vaccines were time consuming. Other physical methods for characterizing the vaccines were unsatisfactory due to the high molecular weight of vaccine particles and their highly negative charge. For example, in Sepharose gel filtration, vaccine samples could only be physically defined by the fact that they were a part of the void volume.

Hydrophilic polystyrene particles (carboxylate-modified) of 45 and 46.5 nm radius (Interfacial Dynamics Corp., Portland, OR 972070) were used as size standards. Sample preparation for electrophoresis and rationales for choosing the polystyrene particles were described in Sections 2.1 and 2.2 of [5].

Development of computer-assisted 2-D agarose electrophoretic procedures

In search of better analytical methods, a computer-assisted electrophoretic method for analyzing such vaccines was developed. Since the vaccine particles were very large with sizes in the range of intact viruses, dilute agarose gels were the medium of choice. However, one-dimensional gels yielded only an uninformative smear (Fig. 1). By adding a second dimension of electrophoresis using Serwer's horizontal 2-D submarine-type apparatus (Fig. 2), gel patterns could be obtained that varied depending on the conditions of the particular vaccine preparation and were therefore characteristic of each vaccine sample (Fig. 3).

Computer programs ElphoFit [6] and GelFit [7] were employed for the evaluation of the 2-D vaccine patterns. In particular, the program ElphoFit was used to evaluate the gel electrophoretic data and to standardize the gel on the basis of the extended Ogston model [8 - 11], as shown in Fig. 9 of [1]. The output of ElphoFit was then transferred to computer program GelFit, which transformed the original digital images (Fig. 3) from a curvilinear system (Fig. 4) to a rectangular coordinate system of particle radius and free mobility, as shown in Fig. 5. Vaccine samples showed a continuous size distribution over a wide range (polydisperse particle mixture) due to the randomizing steps involved in the vaccine preparation (sonication and crosslinking). Another application of GelFit was the stripping of 2-D Gaussian surfaces (Fig. 6), which was used to demonstrate that vaccine II in Figs. 3 and 5 consisted of at least three particle populations.

Biomedical significance

The most significant advance made in the reviewed 2-D experiments was that vaccine samples could be characterized according to particle size and free mobility (related to surface net charge density), even though the immunogens exhibited a polydisperse size distribution with particles as large as or larger than intact viruses. 2-D vaccine patterns varied depending on the conditions of a particular vaccine preparation and, therefore, they were characteristic for each conjugate preparation. The electrophoretic characterization may also be useful for determining vaccine effectiveness and the effects of procedures such as storage, lyophilization, and sterile filtration on the composition of the vaccines. Other more general biomedical applications are the analysis of complex mixtures of macromolecules, small cell organelles, viruses, or other particles of sub-cellular size.

Acknowledgements

The author thanks the late Andreas Chrambach (Section on Macromolecular Analysis, NICHD, NIH) for support, stimulating discussions and for providing laboratory space from 1983 - 93, Akram Aldroubi and Michael Unser (Biomedical Engineering and Instrumentation Program, NIH) for supplying computer program GelFit, and Benes Trus (Center for Information Technology, NIH) for instruction in image processing. Philip Serwer (University of Texas Health Science Center, San Antonio, TX) provided training in the use of the 2-D electrophoresis apparatus in his laboratory. Vaccine samples were kindly provided by Rachel Schneerson (Laboratory of Developmental and Molecular Immunity, NICHD, NIH). Mark Geanacopoulos (JNCI, Oxford University Press, Bethesda, MD) provided a critical review.

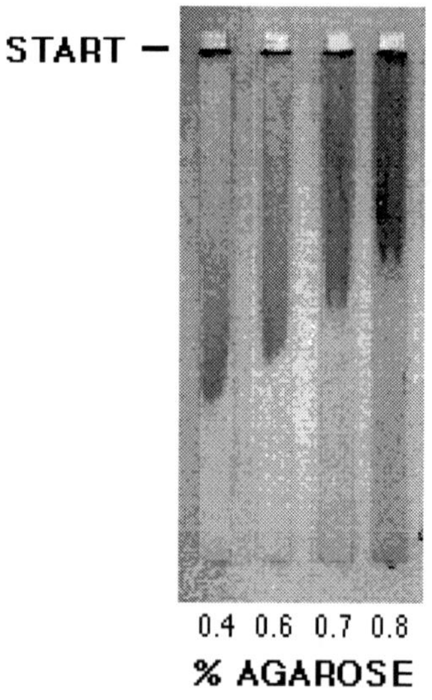

Fig. 1: Gel patterns of meningitis vaccines at different agarose concentrations, using a 1-D submarine electrophoresis apparatus [12]. The samples yielded an uninterpretable smear, although electrophoretic conditions were appropriate.

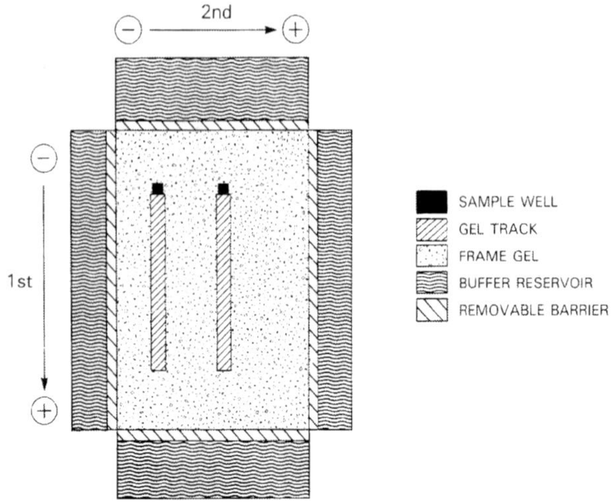

Fig. 2: Schematic of the 2-D horizontal submarine-type apparatus reported by Serwer [13]. Serwer's method, similar to the technique of O'Farrell [14], allows for a separation of particles predominantly according to charge in the first dimension, and according to size in the second dimension. However, the technique of Serwer rests on a different principle: Samples are first electrophoresed in a gel track of low concentration, then the field is switched perpendicularly and the samples are run into a relatively more concentrated frame gel which surrounds the first dimensional track. Gels need not be touched during the procedure; this makes it possible to handle the fragile gels suitable for the separation of nondenatured particles in the size range of viruses (> 3000 kDa). O'Farrell gels, by comparison, are typically used for much smaller proteins or protein fragments (10 - 500 kDa) which are denatured by sodium dodecyl sulfate (SDS). Adapted from Fig. 1 of [1].

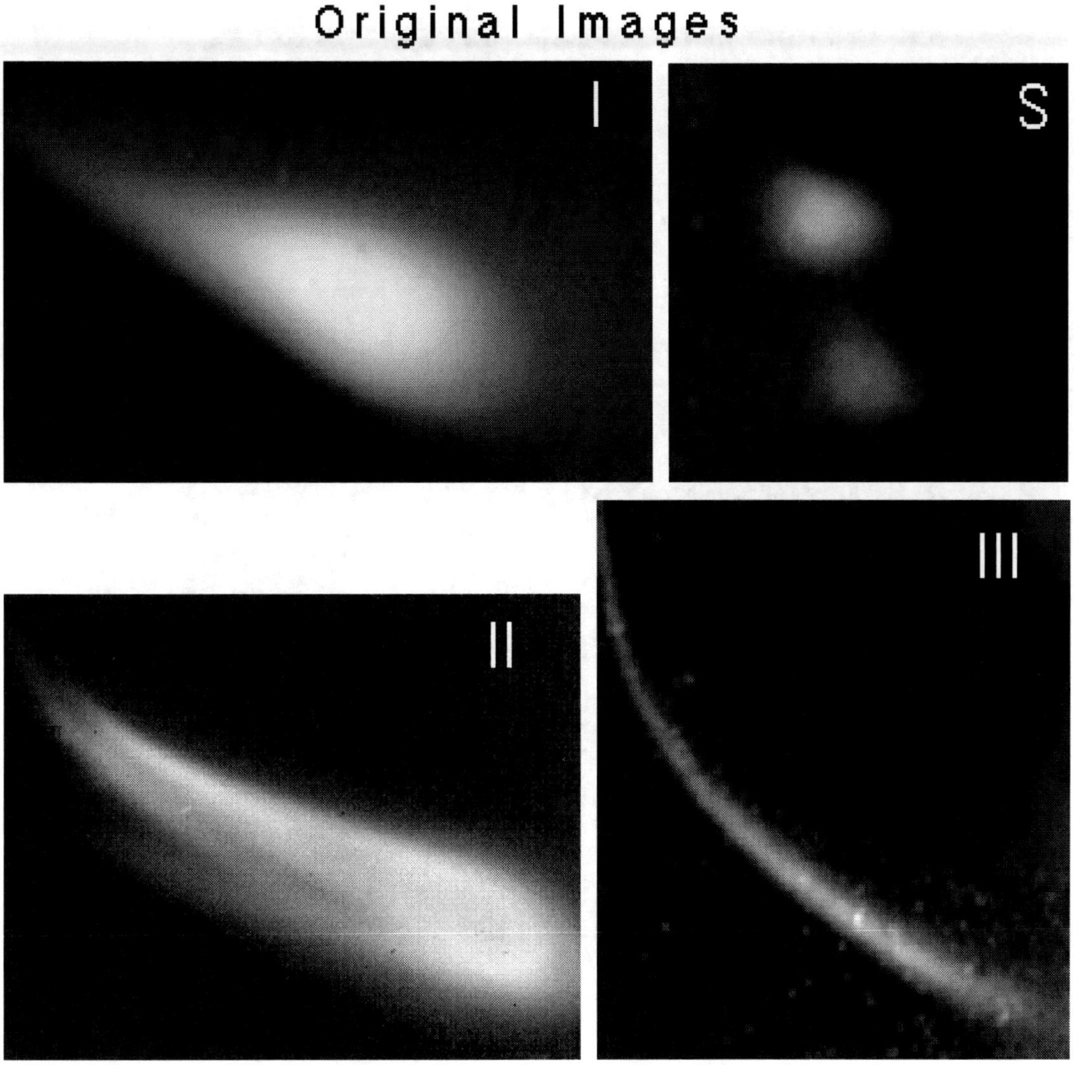

Fig. 3: Two-dimensional gel patterns of meningitis vaccines (I to III) and of hydrophilic polystyrene size standards (S) with 45 and 46.5 nm radius. The origin of electrophoresis lies outside the pictures. First dimension (top to bottom): 0.15 % agarose (SeaPlaque), 3 V/cm, second dimension (left to right): 0.8 % agarose, 1.5 V/cm. Electrophoresis was in phosphate buffer pH 7.2 using the modified [5] 2-D submarine electrophoresis apparatus [13] shown in Fig.2, staining: Coomassie Blue R 250. Scanning of the stained patterns employed a Perkin-Elmer 1010MG microdensitometer interfaced with a PDP 11/34 (Digital) computer. The vaccines produce characteristic gel patterns, which depend on the nature of the sample. The samples consist of *Haemophilus influenzae*, type b, capsular polysaccharide crosslinked to tetanus toxoid (Panels I and II), or P2 protein (Panel III). Experimental conditions and physical parameters of vaccine particles and polystyrene spheres were summarized in Tables 1 and 2 of ref. [15]. (Adapted from Fig. 4 of ref. [1].)

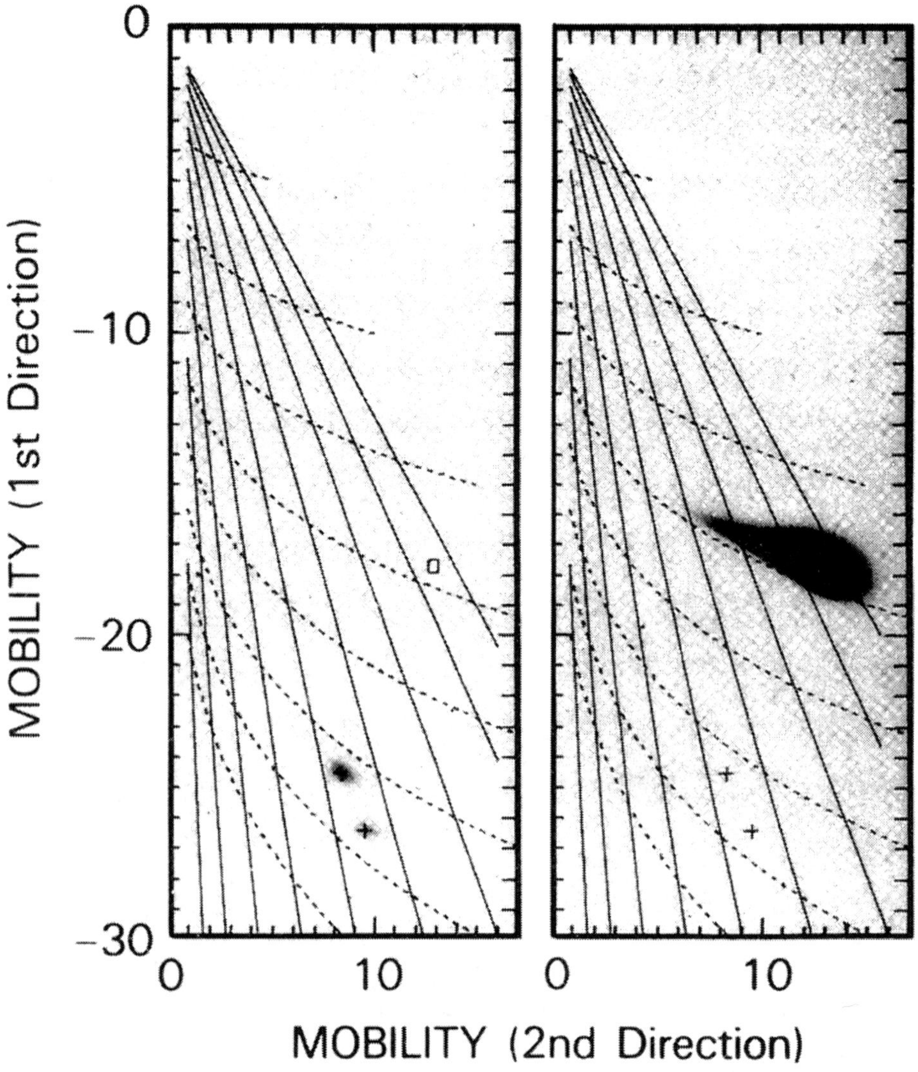

Fig. 4: Iso-size and iso-free-mobility profiles for the evaluation of two-dimensional gel patterns. Hydrophilic polystyrene particles of 45 and 46.5 nm radius are shown on the left, a vaccine sample on the right. Profiles are computed on the basis of the two dimensional positions of co-electrophoresed size standards (left) serving for the determination of gel-specific data. Once these parameters are known, every position on a 2-D gel pattern can be described in terms of particle size and free mobility. Straight solid diagonal lines represent locations of same particle size (10, 20, 30…90 nm, from left to right); curved dotted lines indicate locations of same particle free mobility (5, 10, 15… 40 x $10^{-5}cm^2V^{-1}s^{-1}$, from top to bottom); "+" marks the position of the two standards; the oval the center of the polydisperse vaccine sample distribution. Digitized images of the vaccine sample and the two polystyrene standards are shown in Fig. 3, Panels I and S, respectively. The computation of profiles is briefly described in Section 2.3.2 of [1]; more details are presented in Section 7, Appendix II of [1]. Adapted from Fig. 2 of [1].

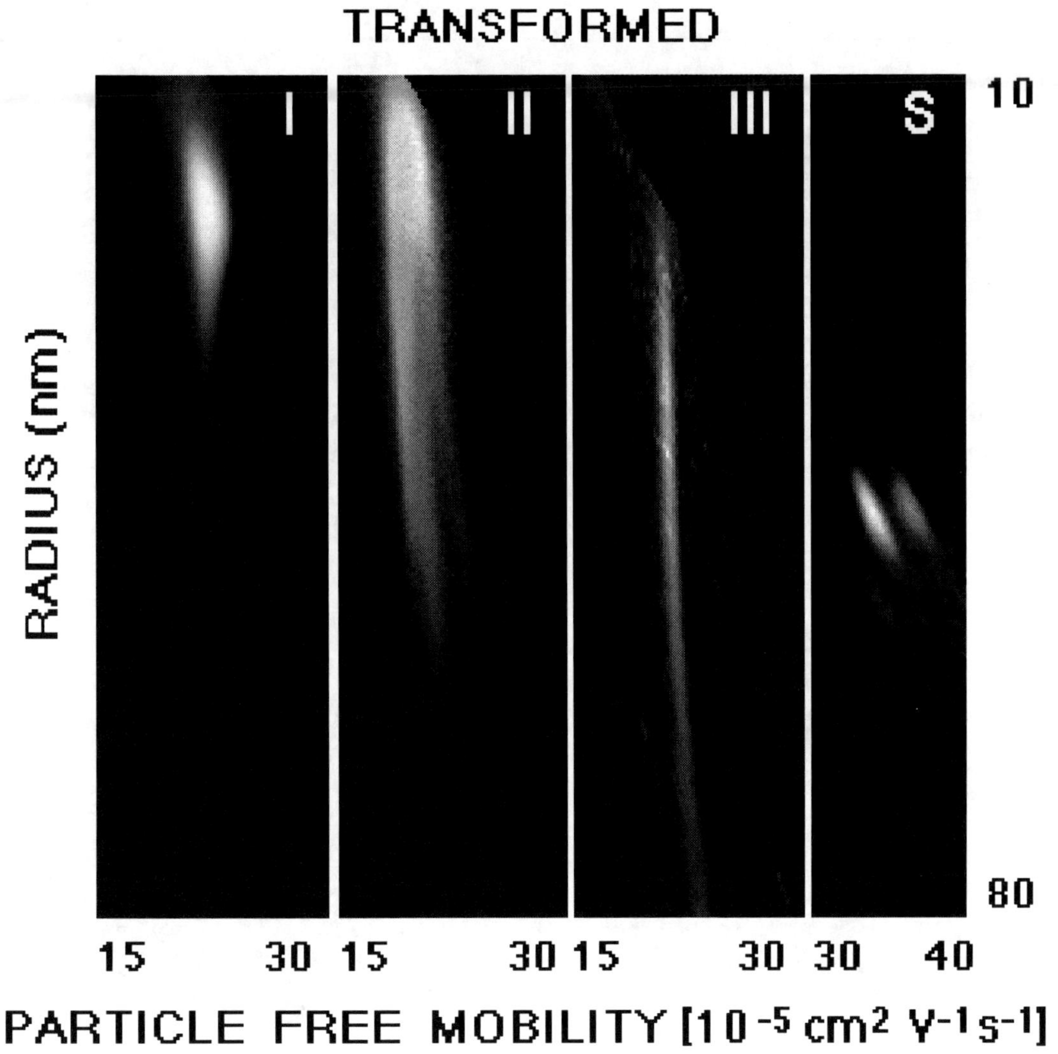

Fig. 5: Patterns of Fig. 3 which have been transformed from the original curvilinear to a rectangular coordinate system of particle size and free mobility (related to surface net charge density). Vaccines II and III, which were effective immunogens have a considerably larger size distribution than sample I (not effective). The vaccines I and II have a much larger variation in free mobility than III, since protein P2 is well defined, whereas tetanus toxoid is a mixture of many components. It should be noted that the 2-D electrophoresis used here achieves results similar to O'Farrell's technique, but relies on a different principle: Predominant charge- (1st dimension) and predominant size-separations (2nd dimension) are achieved by using gels with low and relatively high agarose concentrations under non-denaturing conditions. By applying a mathematical approach (Appendices of [1]) based on the extended Ogston model [8, 9, 10, 11], one can distinguish the separation effects due to particle size and charge. The investigated particles are in the size range of viruses (10,000 to 2,000,000 kDa). Adapted from Fig. 5 of [1]

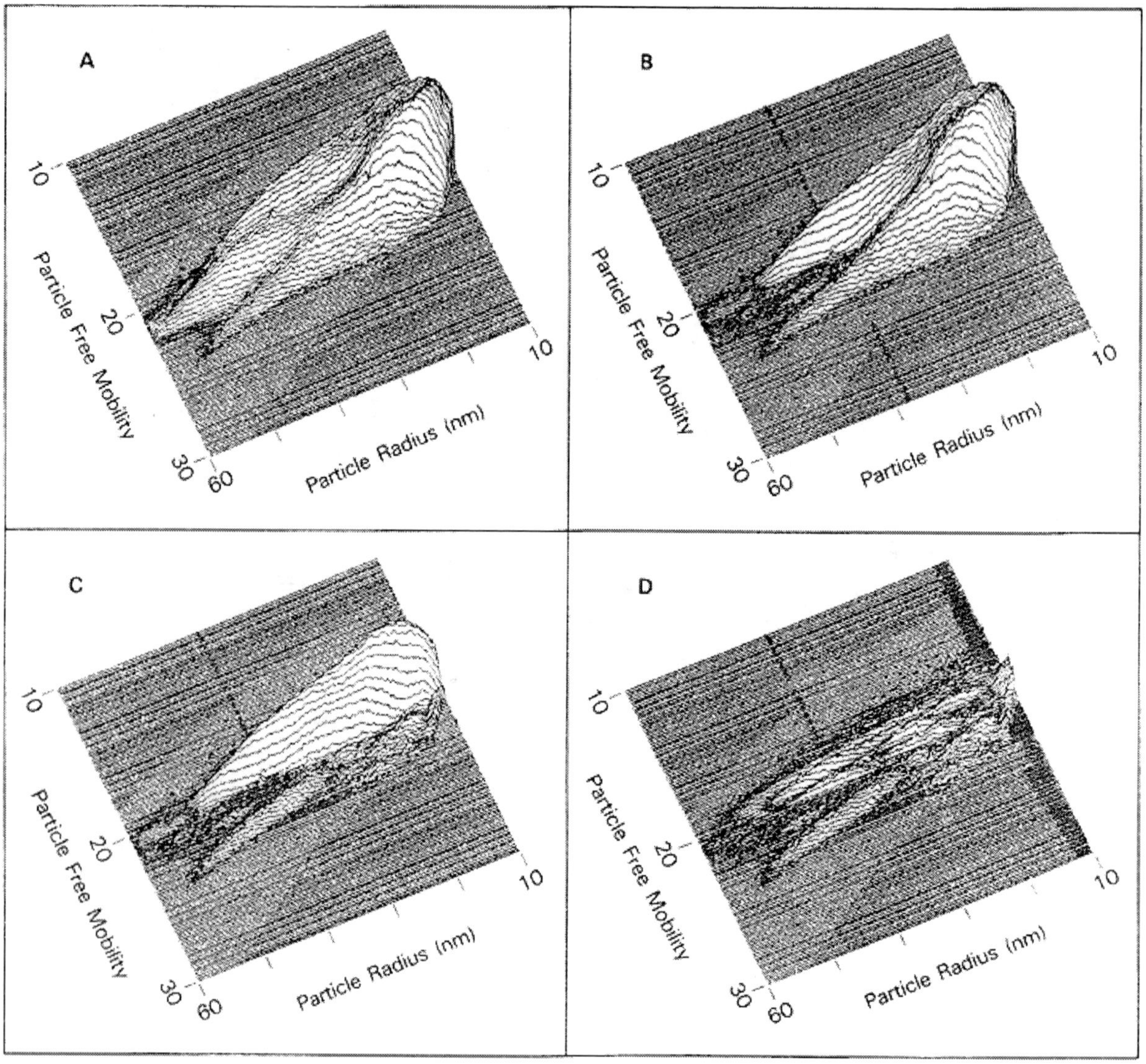

Fig. 6: Progressive stripping of vaccine particle populations (approximated as 2-D Gaussian surfaces) from the transformed image of the vaccine in Fig. 5, Panel II. The distribution of signal energy (relative density) as a function of effective particle radius and free mobility is shown. **(A)** 3-D view of the original transformed image. **(B), (C)** and **(D)** show successive stages of stripping surfaces. It is discussed in Section 3.2.3 of [1] that this particular vaccine sample may be composed of at least three vaccine particle populations. Signal processing by GelFit [7]. 3-D plots were created by program NIH Image, which was developed by Wayne Rasband at the U.S. National Institutes of Health (http://rsb.info.nih.gov/nih-image/). Adapted from Fig. 6 of [1].

References

[1] Tietz, D, Electrophoresis 2007, 28, 512–524.

[2] Schneerson, R., Barrera, O., Sutton, A., and Robbins, J. B., J. Exp. Med. 1980, 152, 361-375.

[3] Claesson, B. A., Trollfors, B., Lagergard, T., Taranger, J., Bryla, D., Otterman, G., Cramton, T., Yang, Y., Reimer, C. B., Robbins, J. B., Schneerson, R., J. Pediatrics 1988, 112, 695-702.

[4] Robbins, J. B., Schneerson, R., Anderson, P., Smith, D. H., JAMA 1996, 276, 1181–1185.

[5] Tietz, D. and Chrambach, A, Electrophoresis 1989, 10, 667-680.

[6] Tietz, D, Electrophoresis 1991, 12, 28-39.

[7] Aldroubi, A., Unser, M., Tietz, D., and Trus, B., Electrophoresis 1991, 12, 39-46.

[8] Ogston, A.G., Trans. Faraday Soc. 1958, 54, 1754-1757.

[9] Rodbard, D. and Chrambach, A., Proc. Natl. Acad. Sci. USA 1970, 65, 970-977.

[10] Tietz, D., Chrambach, A., Anal. Biochem. 1987, 161, 395-411.

[11] Beyer, K., Tietz, D., Adv. Electrophoresis 1988, 2, 145-151.

[12] Serwer, P, Electrophoresis 1983, 4, 227-231.

[13] Serwer, P., Anal. Biochem. 1985, 144, 172-178.

[14] O'Farrell, P. H., J. Biol. Chem. 1975, 250, 4007-4021.

[15] Tietz, D., Aldroubi, A., Schneerson, R., Unser, M., Chrambach, A., Electrophoresis 1991, 12, 46-54.

273d Study of DNA Electrophoresis Mobility as a Function of Its Configuration

Shengnian Wang[1], Xin Hu[2], Orin L. Hemminger[3], and L. James Lee[1]. (1) Department of Chemical and Biomolecular Engineering, The Ohio State University, 140 West 19th Avenue, Columbus, OH 43210, (2) NSEC center for Affordable Nanoengineering of Polymer Biomedical Devices (CANPBD), The Ohio State University, Room 125A, 140 West 19th Ave., Columbus, OH 43210, (3) Chemical and Biomolecular Engineering, Ohio State University, 140w 19th ave, Columbus, OH 43210

Electric fields have been widely used to separate charged DNA molecules in a gel sieving matrix or artificial pillar array within microfluidic devices. It has been shown that the effective electrophoretic (EP) mobility of DNA molecules is dependent on both the size of DNA molecules and the radius of gel matrix or artificial pillar. Large DNA has a decreased EP mobility due to increased collisions with the gel matrix or artificial pillar array. However, the EP mobility of a single DNA molecule in a non-uniform electric field without a gel matrix or artificial pillar is rarely investigated. In this study, electrophoretic DNA dynamics in a straight microchannel and a single microscaled cross-slot were investigated. It was found that the EP mobility of a single DNA molecule is not only dependent on the ion strength of the buffer solution, but also related to the configuration of the DNA molecule and the concentration of surrounding small molecules such as sucrose or glucose. In a zero electric field gradient generated in the straight channel, DNA molecules display a larger value of EP mobility in a high sucrose/glucose concentration solution. While in a constant nonzero electric field gradient produced by single cross-slot, the effective EP mobility of DNA molecule is highly dependent on its aspect ratio. A larger aspect ratio of the DNA molecule leads to a larger effective EP mobility. It is widely accepted that the EP mobility of individual molecules, such as DNA, is constant. However, our findings indicate that the EP mobility of flexible molecules is dependent on the shape of the molecule as it deforms, and the solution it is in. This study provides important fundamentals for the understanding and control of DNA separation with electrophoresis.

273e Clustering Of Counterions On Flexible Polyelectrolytes

Boris Khusid, Chemical Engineering, New Jersey Institute of Technology, University Heights, Newark, NJ 07102, Tak S. Lo, Levich Institute, City College of CUNY, 140th Street & Convent Avenue, New York, NY 10031, and Joel Koplik, Levich Institute and Physics Department, City College of New York, 140th Street and Convent Avenue, New York, NY 10031.

Polyelectrolytes consist of repeating ionizable groups which dissociate in water or other polar solvents, making the macromolecules ionized. Natural and synthetic polyelectrolytes are widespread in applications, ranging from detergents and water treatment to proteomics and to signal transduction in cells. While each of these applications is independently unique, all of them share the common feature of utilizing intermolecular interactions in a polyelectrolyte solution which are governed by the effective charge of a macromolecule being lower than the fully ionized value due to coupling of a macromolecule with surrounding counterions. The effect of the counterion screening of an ionized macromolecule is seen as the critical factor for understanding these complex phenomena. The classic Manning theory of counterion condensation constitutes the basis for thermodynamics of dilute polyelectrolyte solutions (Manning, 1963). The condensed counterions are assumed to be distributed uniformly along the molecule, whereas the unbounded ions in the solution are treated by the Debye-Huckel approximation. The remarkable feature of Manning's model is the independence of counterion condensation on the diameters of the counterions and the polyelectrolyte, as well as the universal nature of the effective charge of the polyelectrolyte above the condensation transition.

Manning's concept of counterion condensation was extended by numerous analytical investigations within mean-field and strong-coupling approximations as well as extensive Monte Carlo and molecular dynamics (MD) simulations. While these studies provided a good understanding of the effects of the counterion screening on the thermodynamic properties of polyelectrolytes, little is known about the dynamics of counterion– polyelectrolyte association and spatial and temporal variations of the counterion distribution around its backbone. Besides the fundamental interest in this problem, strong counterion- polyelectrolyte correlations drastically affect local electrostatic interactions that govern phenomena which occur at such time and length scales, like docking of small molecules to biological macromolecules, assembly of polymeric and colloidal materials, polymer-surfactant interactions, and transport in and out of cells.

Here we consider mutual approaching and parting of counterions and a meandering intrinsically flexible polyelectrolyte. In order to address the electrostatic interaction between a flexible polyelectrolyte and the counterions at both large and small length scales, our MD studies were conducted over a broad range of length scales from a few Debye lengths down to the Bjerrum length and to the ionic diameter. Such simulations require details of the charge distribution across the polyelectrolyte at the scale of the ionic diameter and properly addressing the excluded volume effects of ions and neutral solvent molecules as well as their hydrodynamic interactions. In this regard, our simulations employed a model of a flexible polyelectrolyte suspended in a solvent consisting of explicit solvent atoms, with or without added salt.

Using MD simulations, we investigate dynamical clustering of such counterions while drifting together with the polyion as the degree of its ionization varies from well below to well above the condensation transition. We observe that strong electrostatic correlations cause counterions coupled with the polyelectrolyte conformation to form clusters within which they are packed with short range orientational order with respect to one another. These clusters randomly grow and shrink in length due to counterions incoming and leaving the polyelectrolyte. The motion of the counterions toward and away from the polyelectrolyte is found to be uncorrelated. Surprisingly, the peculiar trends of counterion dynamical clustering appear to be robust to changes in the degree of the polyelectrolyte ionization as well as to the field driven polyelectrolyte electrophoresis.

Our results show that “counterion condensation" is a dynamic process in which individual ions enter and leave the vicinity of the polyelectrolyte independently and randomly on a time scale comparable to the diffusion time and that Manning's picture is correct only on the long time average. On the shorter time scale, the amount of charge in the vicinity of the molecule is wildly fluctuating. We analyzed the time fluctuations of the ions by calculating the average time correlation of the net ionic charge within the Bjerrum and the Debye layers. The electrostatic interaction between counterions and a flexible polyelectrolyte at the local level on the fast time scale is found to be closer to the classic Bjerrum model for the formation of dipole-like structures by two oppositely charged ions in simple electrolyte solutions.

Our simulations provide information on the detail arrangement of the condensed counterions along a flexible polyelectrolyte. Next we investigate the motion of the polyelectrolyte and the small ions in a constant DC electric field. From the simulation data, we extract the electrophoretic mobilities of the polyelectrolyte, the counterions and the coions. These results are compared with predictions of theories for the electrophoresis and conductivity of dilute polyelectrolyte solutions.

273f Nanoparticle-Composite Gels For Protein Separation: Synthesis And Preliminary Characterization

Hope E. Sedrick[1], Jennifer R. Bollig[1], Holly A. Stretz[2], and Pedro E. Arce[3]. (1) Tennessee Technological University, Cookeville, TN 38505, (2) Department of Chemical Engineering, Tennessee Technological University, Prescott Hall, Box 5013, Cookeville, TN 38501, (3) Chemical Engineering, Tennessee Technological University, Department of Chemical Engineering, P.O. Box 5013, Cookeville, TN 38505

Currently, there is an interest in novel drug delivery systems, diagnostic capabilities and improving separation of biomacromolecules such as DNA and proteins. After the successful modification of the gel morphology by using DNA, Xanthan, and SDS (1) as templating agents, electrophoresis efficiency for protein separation increases. Motivated by this success, we are currently using charged nanoparticles to modify the internal gel architecture as well as manipulate the electrostatics properties of the gels. Recently (2), the addition of nanoparticles has shown a modification of the electrosmotic flow. We believe that both the electrosmosis and the gel morphology will affect tremendously the separation efficiency. In this project, polyacrylamide gels were successfully cast and crosslinked with well dispersed, charged nanoparticles of varying diameters (Southern Clay Laponite RD and an experimental Laponite) at a concentration of approximately 1% (w/w). The dispersion of the nanoparticles, or filler, is characterized by the visual clarity of the resultant gels and will be further characterized by acoustical testing, and possibly other microscopic techniques. In this presentation, the authors will discus the synthesis and preliminary characterization protocols for the novel nanoparticle composite gels. Preliminary electrophoresis runs comparing the novel gels with standard gels used in current electrophoresis separations will be presented. Future work could include, for example, modifying the gel morphology and the electrostatics characteristics to tailor the rate of the macromolecule in applications such as drug delivery systems, bioseparation protocols and other related pharmaceutical applications to optimize the performance of the material.

(1) Rill RL, Locke, BR, Liu, Y, Dharia, J, Van Winkle, D, "Protein electrophoresis in polyacrylamide gels with templated pores," Electrophoresis 17 (1996) 1304-1312

(2) Matos, MA, White LR, Tilton RD. Journal of Colloid and Interface Science 2006 (300): 429-436.

333a Dynamical Analysis of Silver-Staining Gel for 2DE

Yusuke Tatsumi[1], Yoichi Kumada[2], and Michimasa Kishimoto[2]. (1) Kyoto institute of technology, Kyoto, Japan, (2) Department of chemistry and materials technology, Kyoto institute of technology, Sakyouku Matsugasaki Hashigamicho, Kyoto, 606-8585, Japan

The time courses of 2DE gel images were analyzed dynamically by our new system, which consist of the intermittent scanner system and the software for the dynamic analysis of the intermittent gel images. The system can recognize protein spots more elaborately compared with the conventional image analyzing system using a single gel image because our system can recognize the overlapped protein spots by use of sequential gel images that were scanned intermittently in the color developing reaction.

Our software can adjust the gel positions on the sequential images by recognition of artificial color points that were inserted on the 2DE gel just before developing process. Furthermore, the time course of each spot image could be displayed and the detailed information were obtained in addition to the spot volume and/or location. Our noble method can analyze 2DE-gel without the harmful influence of the addition time of stopper on the measurement result, and we can detect the difference of the spot of the cancer and non-cancer tissue protein more precisely than those with the normal approach with the image of single developing time.

333b Quantification And Control Of Passive Flows To Minimize Sample Losses At Channel Junctions

Hui Xu, John S. Paschkewitz, C. Charles Park, and Rajiv Bharadwaj. Microfluidics, Caliper Life Sciences, 605 Fairchild Drive, Mountain View, CA 94043

In the absence of valves, sample material in microfluidic channels can be lost in channel junctions. These losses are strongly affected by "passive" flows that result from an imbalance in the Laplace pressure due to liquid surface curvature in the sample reservoirs, or wells, and hydrostatic pressure resulting from differences in fluid level between wells. We present a systematic study of the behavior of these passive flows using a model microfluidic system and then demonstrate a control strategy to minimize their impact on sample loss. Based on direct observations of liquid menisci in the wells, we found that the surface curvature of the liquid in the well was strongly affected by the contact angle hysteresis and whether or not an electrode pin was inserted. Flow measurements indicated that the passive flow has a time-dependent behavior that may be of importance in microfluidic chip operation. The passive flow effects can be controlled using a pressure compensation strategy that balances the hydrostatic and Laplace pressures. We demonstrate this concept by controlling the passive flow at a channel junction in an isotachophoresis (ITP) chip. The loss of the stacked sample band can be significantly reduced when the passive flow is largely suppressed or directed out of the channel junction and is quantified by computer simulation.

333e Nanoparticle-Composite Gels For Protein Separation: Characterization Based On Acoustic Methods

Hope E. Sedrick[1], Jennifer R. Bollig[1], Holly A. Stretz[2], and Pedro E. Arce[3]. (1) Tennessee Technological University, Cookeville, TN 38505, (2) Department of Chemical Engineering, Tennessee Technological University, Prescott Hall, Box 5013, Cookeville, TN 38501, (3) Chemical Engineering, Tennessee Technological University, Department of Chemical Engineering, P.O. Box 5013, Cookeville, TN 38505

Currently, there is an interest in novel drug delivery systems, diagnostic capabilities and improving separation of biomacromolecules such as DNA and proteins. After the successful modification of the gel morphology by using DNA, Xanthan, and SDS[1] as templating agents, electrophoresis efficiency for protein separation increases. Motivated by this success, we are currently using charged nanoparticles to modify the internal gel architecture as well as manipulate the electrostatics properties of the gels. Recently[2] , the addition of nanoparticles has shown a modification of the electrosmotic flow. We believe that both the electrosmosis and the gel morphology will affect tremendously the separation efficiency. In this project, polyacrylamide gels were successfully cast and crosslinked with well dispersed, charged nanoparticles of varying diameters (Southern Clay Laponite RD and an experimental Laponite) at a concentration of approximately 1% (w/w). In this presentation, the dispersion of the nanoparticles, or filler, will be characterized by acoustical testing. Acoustical testing measures the speed of sound though a material. The methodology is based on the fact that the speed of sound through a material is related to its density and the density of the nanocomposite gels differ from the standard gels used in current electrophoresis separations. Several preliminary results based on the principle described will be presented and analyzed. Other microscopic characterization techniques which ensure the well dispersion of the nanoparticles may also be presented for comparison purposes. Future work could include, for example, modifying the gel morphology and the electrostatics characteristics to tailor the rate of the macromolecule in applications such as drug delivery systems, bioseparation protocols and other related pharmaceutical applications to optimize the performance of the material.

1 Rill RL, Locke, BR, Liu, Y, Dharia, J, Van Winkle, D, "Protein electrophoresis in polyacrylamide gels with templated pores," Electrophoresis 17 (1996) 1304-1312

2 Matos, MA, White LR, Tilton RD, "Electroosmotically enhanced mass transfer through polyacrylamide gels," Journal of Colloid and Interface Science 2006 (300): 429-436.

333f Novel Biased Ac Electroosmosis Micropump Using Symmetrical Electrode Array

Nazmul Islam, ECE Dept., Northern Arizona University, Flagstaff, AZ 86011 and Jie Wu, Dept. of Electrical and Computer Engr., The University of Tennessee, Knoxville, 420 Ferris Hall, 1508 Middle Drive, Knoxville, TN TN 37996.

Abstract:

The interaction of AC electric field with the double layer surface charge creates uni-directional force, which in turn produce fluid flow for AC electroosmosis (ACEO). In the last decade ACEO has received increasing interest due to its important applications in micro total analysis systems and miniaturized biomedical devices. With the application of very low voltage (~$1V_{rms}$), ACEO can be utilized to control and manipulate particles and fluids at the micro/nano scale, which are very difficult to achieve with existing techniques, such as pressure driven flow.

A novel micropump is presented in this research for lab-on-a-chip applications that can be fabricated with MEMS-compatible semiconductor micro-fabrication. The proposed AC electroosmotic micropump is based on biased ACEO technique for symmetrical electrode array. Biased ACEO is realized by applying biased AC signals over electrode pairs, leaving the electrolyte floating; therefore, two electrodes have different electrical potentials relative to the electrolyte. With a biased AC signal, $V_{applied}=V_0(1+\cos\omega t)$ over the electrodes, the left electrode is always positive and more prone to Faradaic charging, while the other is always negative and subject to capacitive charging.

The main concept of getting the pumping action is to get the uni-directional flow by breaking the reflection-symmetry in the geometry or applied signal. In our research we have broken the symmetry by applying asymmetric voltage on the symmetric electrode pattern, which eventually breaks the symmetry of the pattern. The corresponding numerical simulations using the finite element software Comsol multiphysics (FEMLab) are also conducted to verify the concept. The good agreement between the simulations and the experimental data regarding the uni-directional flow is also demonstrated. Our analysis also has shown that by decreasing the micro-channel height, the higher surface-to-volume ratio can be achieved, that eventually increases the micro-pumping velocity. At the end of the paper feedback control of micropump is also presented.

Keyword: AC electroosmosis, micropump, microfluidic, symmetric electrode, microflow, micro-fabrication

EO
+
+
-
Positively biased electrodes
Negatively biased electrodes
L_{array}
L_{pair}
+
Net fluid Motion

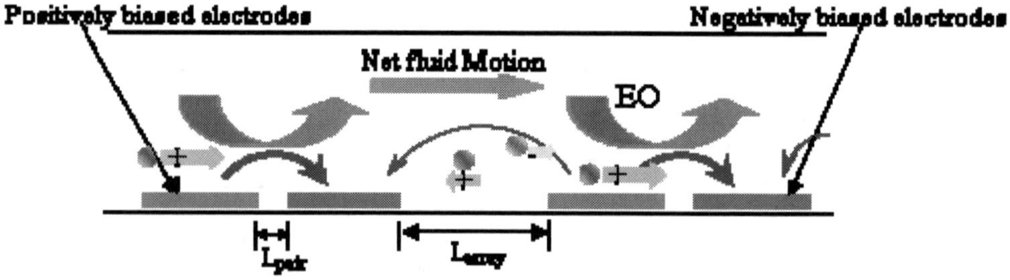

Figure: Biased ACEO produce uni-direction fuid motion, which also imparts differential velocities to particles with various charge/mass ratio.

333g Rapid Cell Rupture In A Dielectrophoretic Field

Adrienne R. Minerick, Sheena Reeves, Ashley Pate, and Sarah Thompson. Dave C. Swalm School of Chemical Engineering, Mississippi State University, Box 9595, Mississippi State, MS 39762

Laboratories-on-a-chip and micro Total Analytical Systems (microTAS) are seen as one of the key growth industries for the 21st century. These systems are attractive due to the promise of raw unprocessed samples entering the device with quantitative analysis results as an output from the device. Because of their small size, these microdevices require small sample volumes and conceivable can return quantitative results within a matter of minutes. Such devices also have the potential to decrease the cost of analysis for applications such as medical testing, do not require technician operation, and can be adapted as point of care devices to diagnose and monitor diseases.

Electrokinetic tools show promise in the manipulation and subsequent quantification of analytes within microdevices. This work examines the use of special electric fields to reliably and rapidly rupture erythrocytes, which can be useful for subsequent subcellular analysis. Rupturing frequencies around 1 kHz were utilized and the field intensity for a series of dependence experiments was varied from 1Vpp/200?m to 10Vpp/200?m. All eight blood types were screened first as fresh samples, then as a function of age. Cell counts were tabulated as a function of time, treatment frequency, amplitude, and age. Results from this process show that there exists a relationship between the age of the blood sample and the applied field.

Web Page: www.che.msstate.edu/research/MDERL/

333h Modeling Of High Frequency Diamond Layered Saw Devices For Potential Microfluidic Applications

Subramanian Sankaranarayanan[1], Stefan Cular[2], and Venkat R. Bhethanabotla[2]. (1) University of South Florida, Dept. of Chemical Engg., ENB 118, 4202 E Fowler Avenue, Tampa, FL 33620, (2) Department of Chemical Engineering, University of South Florida, 4202 East Fowler Ave., ENB 118, Tampa, FL 33620

Abstract

Integration of microelectronics and biotechnology requires developing interfaces that are compatible with microelectronics processing methods, and also provide the requisite selectivity and stability when exposed to biological environments. Diamond is especially attractive because, in addition to having good electrical and chemical properties, it is widely considered to be biocompatible, and can be used as a waveguide in layered SAW devices. Additionally, the high density of diamond layer helps achieve higher acoustic wave propagation velocity and realize GHz frequency devices with transducer finger spacing achievable by optical lithography. In the present work, we use 3-D structural and 3-D/2-D fluid solid interaction finite element models to study the wave propagation characteristics in diamond layered piezoelectric surface acoustic wave devices under the influence of a fluid loading.

The structural model is first utilized to study the wave propagation in diamond layered SAW devices. The 3-D FE model describes two-port structures on the surface of LiNbO3 and LiTaO3 substrates. The simulated device consists of three finger pairs in each port. The fingers are considered as mass-less electrodes to ignore the second-order effects arising from electrode mass, thereby simplifying computation. The periodicity of the finger pairs is 40 microns and the aperture width is 200 microns. The transmitting and receiving IDT's are spaced 130 microns or 3.25λ apart. The simulated SAW device dimensions are 800 microns in propagation length, 500 microns wide and 400 microns deep. The simulated models have a total of approx. 250000 nodes and are solved for four degrees of freedom (three displacements and voltage). The model was created to have the highest densities throughout the surface and middle of the substrate. The simulated voltage and displacement profiles obtained at the output IDT's are used to measure wave attenuation and velocity changes and thereby optimize the thickness of the waveguide, i.e. the diamond layer. The optimized waveguide thickness is then used in the fluid-solid interaction FE model to investigate the influence of fluid loading on Rayleigh and SH wave propagation in diamond layered SAW devices and evaluate their use as potential biosensors.

333i Predicting Electrohydrodynamic Flow Rates In Capillaries To Model Transdermal Drug Delivery

Jennifer Anne Pascal, Chemical Engineering, Tennessee Tech University, PH-214, Cookeville, TN 38505, Mario A. Oyanader, Universidad Catolica del Norte, Avenida Angamos, Antofagasta, 0610, Chile, and Pedro E. Arce, Chemical Engineering, Tennessee Technological University, Department of Chemical Engineering, P.O. Box 5013, Cookeville, TN 38505.

Electrokinetic-based methods to deliver drugs through the skin are becoming more attractive as many medicines are potentially harmful to the digestive system and provide only temporary relief. Many other methods of drug delivery have been studied, including aerosol sprays for the lungs as well as patches for the skin. Electrokinetic-based methods are being studied in order to model drug delivery through the skin to ultimately assist in the design of procedures to deliver drugs via the skin. This project involves using idealized geometries (rectangular, cylindrical, and annular) of a capillary bundle to model the skin. The use of hydrodynamic and electrostatic principles allows the analysis of fluid flow within the capillaries as well as the transport of the drug. Predicting flows in capillaries may be accomplished by applying the principles of electrokinetics in conjunction with hydrodynamics. In this project, capillaries of rectangular and cylindrical geometries are examined, for which velocity profiles are determined and parametrically studied. Afterwards, the flow rates are computed. In order to gain a deeper understanding of how these predictions differ according to the geometries used, the two situations described below are studied.

First, a ratio of the side magnitudes of the cross-sectional area of the rectangular geometry is selected in order to yield analytical velocity profiles. Specifically, velocity profiles with a 1D spatial dependence are described. This allows for an effective comparison of the flow rates predicted by the two different geometries. In fact, the ratio of the electrokinetic contribution of the rectangular geometry to the cylindrical geometry is placed between two limits whose values depend on the inverse Debye length. For low values of this parameter, the pure hydrodynamic limit is reproduced while, for large value of the Debye length, a lower value is obtained.

Second, and in order to see how the limits identified for the 1D case (see above) are affected, 2D spatial-dependence velocity profiles are identified. This case is solved numerically for the velocity profiles and values of the flow rate are afterwards computed.

The presentation will describe details of the two cases identified above. The analysis will be illustrated with dimensionless functions of the volumetric flow rates for both geometries, and quantitative comparisons of the prediction of the two analyses will be discussed. These flow rates are useful to predict the optimal time for many applications, including drug delivery and separations and assess the potential error made by choosing a given geometry over another.

333j A Simplified Mathematical Model Of An Ek Cell To Assess The Influence Of Buoyancy Driven Flows On Removal Efficiency

Cynthia Torres, Chemical Engineering, Tennessee Tech University, PH-214-Chemical Engineering, Cookeville, TN 38501, Mario A. Oyanader, Universidad Catolica del Norte, Avenida Angamos, Antofagasta, 0610, Chile, and Pedro E. Arce, Chemical Engineering, Tennessee Technological University, Department of Chemical Engineering, P.O. Box 5013, Cookeville, TN 38505.

Soil clean up embraces a large family of approaches based on the fundamental principles applied in the operation. For example, EK soil remediation, identified as a versatile technology that can be applied to heavy metal as well as hydrocarbons, has a family of at least seven sub-technologies (i.e. Cation Selective Membrane, Ceramic Casting, Lasagna, Electrochemical ion exchange, etc). Although temperature development on the soil matrix has been observed, little effort has been dedicated to demonstrate or document how temperature triggered mechanisms may be used to overcome soil remediation pitfalls. This contribution focuses on the analysis of results obtained by simulation with a simplified mathematical model. The general hypothesis is that there exist an effect of buoyancy driven flows on the over all removal efficiency of contaminant and that the inclination a EK cell may be used to enhance solute transport. To validate the findings, an experimental approach will be outlined. On the overall, the results and conclusions derived from this study are the baseline for the EK cell system modeling. Comprehensive details of key factors are highlighted to illustrate the modeling technique as well as the system representation.

333k Comparison Of Convective-Dispersive Aspects In Electrokinetic Cells Of Rectangular And Cylindrical Geometry

Antonio Martinez[1], Mario A. Oyanader[1], and Pedro E. Arce[2]. (1) Universidad Catolica del Norte, Avenida Angamos, Antofagasta, 0610, Chile, (2) Chemical Engineering, Tennessee Technological University, Department of Chemical Engineering, P.O. Box 5013, Cookeville, TN 38505

The occurrences of flow reversals during EK processes has been explained by Oyanader et al (2003); Oyanader et al. (2005a&b) using different geometrical models for the soil porous matrix. The conjunction of three competing driving forces (pressure, buoyancy and electrostatic) have an affect on the motion of the solute species that need further understanding. The different driving forces need to be studied in term of how they promote mixing or separation of solute species in an EK cell. Most EK technologies, i.e. Cation Selective Membrane, Ceramic Casting, Lasagna, Electrochemical ion exchange, etc., have reported low removal efficiency. In this contribution, the main focus is on unveiling the possible reasons for EK pitfalls from a fundametal point of view. The approach is based on modeling and fundamental principles applied, first, to the hydrodynamic part of the system and, later, to the species/analytes transport. An area averaging technique has been used to capture the main parameters of the system such as effective mixing, and diffusivity. Furthermore, an approach has been applied to describe the capillary aspect of the soil matrix. This approach involves the characterization of the capillary channels as of rectangular and/or cylindrical aspect to present and discuss several limiting cases related to the influence of flow regimes on the transport of solute in an EK for removal purposes. Optimal operating conditions are highlighted to introduce design criteria and protocols.

Oyanader, M., P. Arce, and A. Dzurik, "Avoiding Pitfalls In Electrokinetic Remediation: Robust Design And Operation Criteria Based On First Principles For Maximizing Performance In A Rectangular Geometry", Electrophoresis, 2003. Oyanader, M., P. Arce, And A. Dzurik, "Design Criteria For Soil Cleaning Operations In Electrokinetic Remediation. Hydrodynamic Aspects In A Cylindrical Geometry", Electrophoresis, 2005a. Oyanader, M., P. Arce and A. Dzurik "Design Criteria For Soil Cleaning Operations In Electrokinetic Remediation. Hydrodynamic Aspects In An Annular Geometry", I&ECh Research, 2005b.

333l Effect Of Dispersion Of Ions On The Performance Of An Electrophoretic Cell With Orthogonal Electrical Field

Pablo Vergara, Chemical Engineering and Material Sciences, University Catolica del Norte, Avenida Angamos 0610, Antofagasta, Chile, Mario A. Oyanader, Universidad Catolica del Norte, Avenida Angamos, Antofagasta, 0610, Chile, and Pedro E. Arce, Chemical Engineering, Tennessee Technological University, Department of Chemical Engineering, P.O. Box 5013, Cookeville, TN 38505.

The hydrodynamic of a system, as it has been demonstrated, has an effect on the transport of solute in an electrophoretic cell. The role of an orthogonal field to flow direction on the hydrodynamic of a system has been studied by the authors using a cell of rectangular geometry*. The study has been extended to include an annular system where the role of the radial transport may be different to that in rectangular systems. Furthermore, the role of the cross sectional area size in an annular cell may have a different effect on the movement of ions of different oxidation state. The method of spatial averaging is used in combination with the solute species continuity equation to determine the effect of the disersion of the charged ions on device performance in a Poiseuille type flows, under the stress of electric fields. Explicit analytical expressions are derived for the effective parameters as a function of the applied electric field, capillary porosity and the flow regime. The result of the analysis can be used in the design of separation processes as well as environmental applications such as soil remediation and wastewater treatment.

____________________________ *Mario A. Oyanader and Pedro Arce, Role of Geometrical Dimensions in Electrophoresis Applications with Orthogonal Fields, Electrophoresis 2005, 26, 2857–2866.

333m Role Of Electro-Osmosis Based Flows On The Effective Transport In A Rectangular Geometrical Electrophoretic Cell

Jasna Godoy, University Catolica del Norte, Avda Angamos 0610, Antofagasta, Chile, Mario A. Oyanader, Universidad Catolica del Norte, Avenida Angamos, Antofagasta, 0610, Chile, and Pedro E. Arce, Chemical Engineering, Tennessee Technological University, Department of Chemical Engineering, P.O. Box 5013, Cookeville, TN 38505.

Separation of biomacromolecules embraces a significant number of applications related to, for example, electro-assisted drug delivery, micro-electrophoretic separations, soil remediation processes, and material processing, just to name a few. Although, the influence on the flow regimes on effective transport has been identified in the literature, the specific role of electrosmotic type forces has not been reported. This contribution focuses on the understanding of the role of electro-advective, i.e. electro-osmosis driven flows on dispersion of solutes. Furthermore, this contribution builds the baseline for comparison with different capillary geometries and how separation or mixing may be controlled. Using the spatial averaging method in combination with the solute species continuity equation, the authors have derived analytical expressions for effective parameters, i.e., effective dispersion coefficient and effective convective velocity. Illustrations are presented to demonstrate the effect of different parameters in a capillary channel of rectangular geometry. The influence of applied electric field, surface potential and Debye length on the velocity profiles are examined.

333n Identification And Antibody-Therapeutic Targeting Of Antibiotic-Resistant Outer Membrane Proteins In E. Coli

Hui Li[1], Xiangmin Lin[1], Chuangzhong Huang[1], Sanying Wang[2], and Xuanxian Peng[1]. (1) State Key Laboratory of Biocontrol, School of Life Sciences, Zhongshan University, Guangzhou, 510275, China, (2) Department of Biology, Xiamen University, Xiamen, 361005, China

The elucidation of the molecular details of antibiotic resistance will lead to improvements in extending the efficacy of current antimicrobials. Our research focuses on Om proteins in response to antibiotic resistance. This invesitigation was perforemd by proteomic methodologies for the altered Om proteins from survived bacteria in a suddenly strong antibiotic or in sub-MIC of the drug because bacteria my be exposed in the two treatments in their becoming resisitance to the antibiotic. Their shared proteins were further confirmed by Western blotting and gene mutant methods as potential targets for designation of new drugs to inhibit the growth of the antibiotic-resistant bacteria. Moreover, a novel method of specific antibody combating bacterial growth was developed based on these altered Om proteins. Our results suggests that combination therapy involving antibiotics that enhance the expression of an antibody target could be far more effective than either drug alone. Our study give a novel insight into therapy to infection by antibiotic-resistant bacteria. Our study has been extended into a variety of antibiotics including tetracycline, ampicilin, chloramphenicol, chlortetracycline, Streptomycin, kanamycine, nalidixic acid, ceftazidime, ciprofloxacin, balofloxacin, ceftriaxone sodium. *Corresponding author: Dr. Xuanxian Peng This work was sponsored by grants from Guangzhou Key Project 2006Z3-E0251

333o Rapid Screen Of Highly Efficient Vaccine Candidates By Immunoproteomics And Cross Immunoproteomics

Xuanxian Peng, State Key Laboratory of Biocontrol, School of Life Sciences, Zhongshan University, Guangzhou, 510275, China and Sanying Wang, Department of Biology, Xiamen University, Xiamen, 361005, China.

Diseases caused by microorganisms can be controlled by vaccines, which require neutralizing antigens. Therefore, it is very important to identify highly efficient immunogens for immuneprevention.

By combining immunoproteomics and bacterial challenge after immunization, we developed a rapid method for screening protected antigens of pathogenic bacteria in aquaculture. Our approach may be divided into three consecutive steps. First, dominant immunogens of outer membrane proteins are screened by immunoproteomics. Second, proteins with the ability to induce production of neutralizing antibodies are identified from the immunogens by virulent bacterium challenge following vaccination. Third, vaccine candidates are determined by evaluation of neutralizing abilities. Information on the candidates has been obtained for further gene cloning by mass spectrometry. Our results indicate that highly efficient protected antigens were identified from the outer membrane proteome of Aeromonas hydrophila, in which an immunogen showed 71.4% protective ability with multivalent functions to A. hydrophila and Aeromonas sobria.

Moreover, a cross immunoproteomics was developed with the use of homogeneous and heterogeneous antibodies as the primary ones, which were prepared from bacterium as an antigen in Western blotting and other bacteria, respectively. These spots recognized by both homogeneous and heterogeneous antibodies were cross-protective antigens and identified with mass spectrometry. Their genes were cloned and expressed. The purified recombinant proteins were applied for investigation of cross-protective immunity in fish and mice. Two highly efficient vaccine candidates determined in V. parahaemolyticus outer membrane proteins could be against infections by not V. parahaemolyticus, but also V. alginolyticus, Aeromonas hydrophila and Pseudomonas aeruginosa, which indicates that two outer membrane proteins show universal function against bacteria.

In summary, the methodologies developed in the current study is high throughout and hence approached rapid, highly efficient and accuracy aims, which will play active role in immune prevention for microbiological diseases. This work was sponsored by grants from National Basic Research Program of China (2006CB101807) and “863” project (2006AA09Z43Z).

337a Electrophoresis Of DNA Molecules Confined In Virtual Nanochannels

Gary W. Slater and Martin Bertrand. Physics, University of Ottawa, 150 Louis-Pasteur, Ottawa, ON K1N 6N5, Canada

We present the first theoretical and computational study of polyelectrolytes undergoing electrophoresis in free solution while being squeezed by a lateral mechanical force (a harmonic potential with a cylindrical geometry) transverse to the direction of the electric field. We show that the polylectrolyte retains its “free-draining” properties in such situations, a result which brings a new perspective on a most fundamental property of charged biopolymers. The electrophoretic mobility of the chains can be described in terms of a local model where the relevant effective segment is oriented along the direction of motion. This description predicts an increase in mobility as the squeezing potential is strengthened, in agreement with our Molecular Dynamics simulation data. Interestingly, our data also suggest a possible novel separation scheme. This is quite different from the previously examined case where a mechanical force is applied at the extremity of a polyelectrolyte in the direction of the its net motion (as in End-Labelled Free-Solution Electrophoresis, or ELFSE) since such longitudinal forces are known to affect the free-draining behaviour.

337b Molecular Dyamics Simulations Of Dynamic Coatings For Quenching Electroosmotic Flow

Owen A. Hickey and Gary W. Slater. Physics, University of Ottawa, 150 Louis Pasteur, Ottawa, ON K1N6N5, Canada

The suppression of electroosmotic flow (EOF) through the use of a dynamically adsorbed polymer coating is a widely used technique in micro- and nano-fluidics, particularly for capillary electrophoresis. Recent experimental evidence has suggested that the most effective coatings are those which are only weakly adsorbed to the surface. This is because the suppression of the EOF depends on the amount of polymer which stretches from the surface through the debye layer and into the bulk and not the friction felt between monomers and the surface. In this talk we report molecular dynamics simulation where we vary the strength with which polymers are adsorbed to the surface of a small tube. Our results show that the optimal adsorption strength for the suppresion of EOF is around the phase transition for adsorption of the polymer. In the case of very strongly adsorbed polymer coatings, the coating can actually increase the EOF by thickening the debye layer since it reduces the effective radius of the capillary. Simulation results also show that at very high electric fields the adsorbed polymer layer is deformed reducing the effectiveness of the coating through a reduction in the height of the adsorbed polymer layer. Finally it is shown that very short chains make for ineffective coatings since they are not long enough to loop from the wall into the bulk of the fluid.

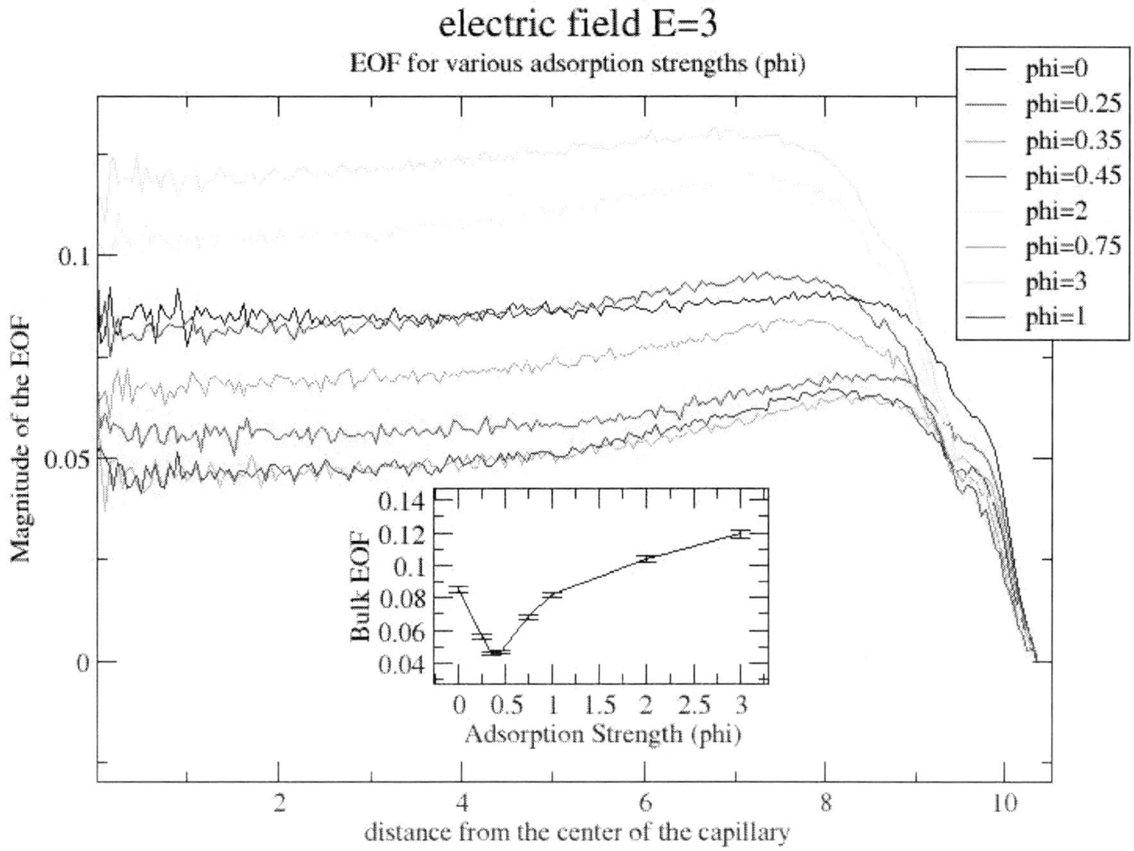

337c The Nanoporous Morphology Of Hydrogel Separation Matrices For DNA Electrophoresis

Jian Wang[1], Hugo Y. Acosta[2], and Victor M. Ugaz[1]. (1) Artie McFerrin Department of Chemical Engineering, Texas A&M University, 3122 TAMU, College Station, TX 77843, (2) Department of Chemical Engineering, Universidad Industrial de Santander, Bucaramanga, Colombia

Polymer hydrogels play a key role as sieving matrix materials for electrophoresis of charged biomolecules like DNA and proteins. In order to rationally select hydrogels that possess properties optimally suited for these applications, it is important to understand the nature of the gel pore structure and the parameters through which it can be manipulated. The characteristic average pore size and degree of polydispersity are particularly important quantities but they have proven difficult to measure experimentally because most hydrogels are mechanically fragile and easily damaged. We have recently developed an approach that uniquely combines complementary information from in-situ dynamic mechanical analysis, thermoporometry, and transmission electron microscopy in a way that enables both the mean pore size and the pore size distribution to be quantitatively characterized in polymer hydrogels. Here, we apply these techniques to investigate how chemical composition (relative amounts of monomer and crosslinker) and polymerization conditions (chemical versus UV initiation chemistry) influence the nanoscale pore morphology in polyacrylamide/bis-acrylaimde hydrogels. When compared with corresponding measurements in agarose gels, these measurements provide insights into the interplay between the nanoporous architecture of the gel matrix and electrophoretic separation behavior. This information also allows us to identify synthesis conditions that yield hydrogels with comparable mean pore sizes, but whose pore size distributions incorporate different degrees of polydispersity. The critical role played by polydispersity becomes evident when we examine controlled release of macromolecules from these gels, where we observe dramatically altered release kinetics in gels with different polydispersity that would otherwise appear identical if compared on the basis of their mean pore size alone.

337d Engineered Entangled Polymer Networks And System Design For Improved Four-Color Sequencing Of DNA By Microchip Electrophoresis

Daniel G. Hert, Christopher P. Fredlake, and Annelise E. Barron. Department of Chemical and Biological Engineering, Northwestern University, 2145 Sheridan Road, Evanston, IL 60208

Entangled polymer networks continue to be of great interest as a means to separate single-stranded DNA fragments, particularly for DNA sequencing applications. While capillary electrophoresis instruments were very successful in sequencing the complete human genome as part of the Human Genome Project, the capital costs, required reagent volumes, and long run times make this technology infeasible for continued large-scale de novo sequencing projects. Microfluidic chips have shown tremendous promise for producing high quality DNA sequencing data while also lowering required reagent volumes and separation times. The ability to precisely control the injection plug width in the separation channel by specifying the microchip geometry enables shorter separation channels to be used, which result in shorter separation times. Performing DNA sequencing fragment separations in microchips, our group has previously described a poly(N,N-dimethylacrylamide) (pDMA) matrix and physically adsorbed poly (N-hydroxyethylacrylamide) wall coating system that enables ultra-fast DNA sequencing of 600 bases in approximately 6.5 minutes at 98.5% accuracy. This result is a 66% decrease in separation time over previously published results. To further explore this observed improvement in separation speed, we have varied pDMA matrix properties, such as polymer molar mass and polymer solution concentration, as well as several chip electrophoresis parameters, particularly separation length, temperature, and electric field strength, to determine the resulting sequencing read lengths and separation times. In this report, we discuss the effect each of these parameters has on both sequencing read length and separation time. We also discuss the effect that blending two pDMA samples of differing molar mass has on the resulting sequencing read lengths and separation time in microfluidic channels. These results will also be discussed in the context of different sequencing applications, specifically medical sequencing and full genome sequencing.

337e Microfabricated Microemulsion Droplet Generator Enables The Pcr Colony Front End Of The Microbead Integrated DNA Sequencer

Palani Kumaresan[1], Chaoyong James Yang[2], and Richard A. Mathies[2]. (1) Department of Mechanical Engineering, University of California, Berkeley, Berkeley, CA 94720, (2) Department of Chemistry, University of California, Berkeley, Berkeley, CA 94720

The Microbead INtegrated DNA Sequencer (MINDS) project is advancing Sanger sequencing technology to its ultimate limit: the production of complete sequencing separations from a single template PCR colony by coupling bead-based PCR template amplification with integrated microfluidic extension and separation. As a part of this process we have recently shown that Sanger extension and sequencing separation can be effectively integrated by the use of an oligonucleotide affinity gel matrix for extension sample purification and concentration [1]. We have further demonstrated high quality sequencing reads from only 100 attomoles of PCR amplicon in a sample purification-CE microdevice that incorporates a more efficient inline injection system [2]. These advances now make it feasible to sequence directly from the PCR products generated by a single molecule in controlled size emulsions. To enable a high-throughput front end for attomole-scale Sanger DNA sequencing and other genetic analysis, we have now developed a microfabricated droplet generator [3] to rapidly form highly monodisperse, nanoliter volume droplets of PCR mix in emulsion oil. Around 2000, 5 nL droplets, formed this way, are collected in individual tubes and simultaneously temperature cycled using a block thermal cycler. The generator also provides a reliable way to introduce primer functionalized microbeads in the droplets, which act as substrates for solid phase PCR and facilitate extraction of amplified products from individual droplets. To evaluate this system, a 545 bp region of the pUC18 genome was successfully amplified on ~34 μm diameter beads contained in ~5 nL volume droplets. Sufficient sequencing template can be amplified from single DNA molecules on individual beads to enable Sanger extension and sequencing in the ultra-efficient integrated sequencing processors [1, 2]. The work presented here establishes the feasibility of the MINDS process that will take Sanger sequencing, which has the unique ability to produce long and accurate sequence reads, to its fundamental molecular limit.

REFERENCES

[1] "Microfabricated bioprocessor for integrated nanoliter-scale Sanger DNA sequencing", R.G. Blazej, P. Kumaresan, R.A. Mathies, Proc. Natl. Acad. Sci. USA, 103, 7240 (2006).

[2] "Inline-injection microdevice for attomole-scale Sanger DNA sequencing", R.G. Blazej, P. Kumaresan, S.A. Cronier, R.A. Mathies, Anal. Chem., In press.

[3] "High-temperature microfluidic synthesis of CdSe nanocrystals in nanoliter droplets", E.M. Chan, A. P. Alivosatos, R.A. Mathies, J. Am. Chem. Soc., 127, 13854 (2005).

337f Microfluidic Selection Of Aptamers

Michael T. Bowser, Bryan R. Fonslow, Renee K. Mosing, and Ryan Turgeon. Chemistry, University of Minnesota, 207 Pleasant St. SE, Minneapolis, MN 55455

Aptamers are single stranded DNA or RNA molecules that fold into a secondary structure that generates high affinity and selectivity for particular target molecules. Aptamers have gained widespread attention for their potential as drug or diagnostic candidates. Unfortunately, the process for isolating aptamers remains technically challenging and time consuming. We have made advances in these areas with the introduction of capillary electrophoresis based aptamer selections (CE-SELEX) but room for further improvement remains.

In micro free flow electrophoresis (μFFE), a thin stream of sample is continuously introduced into a planar, microfluidic flow chamber. An electric field is applied perpendicularly to the flow through the separation chamber. Analytes are deflected laterally in the electric field according to their electrophoretic mobility. Preparative separations can be performed by continuously collecting fractions at the exit of the flow chamber.

μFFE is in many respects an ideal separation technique for isolating binding aptamers from large random sequence nucleic acid libraries. The target is incubated with the library and pumped into the separation chamber. Binding sequences are separated from nonbinders laterally in the separation chamber and collected as a separate fraction. The continuous nature of the separation allows large numbers of sequences to be separated without using extreme library concentrations. The continuous separation also eliminates the precise timing required for successful fraction collection in CE-SELEX.

396a Immunomagnetic Cell Separation: Fundamentals, Current Issues, and Future Potentials

Jeffrey Chalmers, Chemical and Biomolecular Engineering, Ohio State University, 140 W. 19th Avenue, Columbus, OH 43210

Magnetic separation technology (both cell and molecular) has grown in the last two decades from a research curiosity to a commercial industry which has been estimated to have commercial sales greater than $100 million dollars a year. Fundamental to most magnetic separation technologies is a magnetic force, be it attractive (i.e. paramagnetic, superparamagnetic, ferromagnetic) or repulsive (i.e. diamagnetic) that is derived from a gradient of the magnetic field. Consequently, “micro” or “nano” applications are uniquely suited to exploit these small length scales since the magnetic force increases linearly with the increase in a magnetic gradient. Since it is usually desirable to maximize the driving force of any separation technology, one can increase the magnetic force by either increasing the magnetic field, increasing the magnetic gradient, or some combination of both.

This presentation will attempt to summarize/review the fundaments governing and controlling magnetic separation, review current commercial technology with respect to performance capability and limitations, and outline future potentials based on these fundamental principles.

396b Microfluidic Scaffolds for Three-Dimensional Cell Culture and Oxygen Sensing

Nak Won Choi[1], Mario Cabodi[1], Brittany Held[2], Jason P. Gleghorn[3], Hila Dvora[4], Kang-Yeol Park[5], Lawrence J. Bonassar[6], and Abraham D. Stroock[7]. (1) School of Chemical and Biomolecular Engineering, Cornell University, 120 Olin Hall, Ithaca, NY 14853, (2) Chemistry and Chemical Biology, Cornell University, 120 Olin Hall, Ithaca, NY 14853, (3) Biomedical Engineering, Cornell University, 130 Upson Hall, Ithaca, NY 14853, (4) Chemical and Biomolecular Engineering, Cornell University, 120 Olin Hall, Ithaca, NY 14853, (5) 4Samsung Cheil Industries Incorporation, Uiwang-si, South Korea, (6) Sibley School of Mechanical and Aerospace Engineering, Cornell University, 218 Upson Hall, Ithaca, NY 14853, (7) School of Chemical and Biomolecular Engineering, 360 Olin Hall, Cornell Univeristy, Ithaca, NY 14583

Most methods to culture cells in three-dimensions depend on a cell-seedable biomaterial to define the global structure of the culture and the microenvironment of the cells. Efforts to tailor these scaffolds have focused on the chemical and mechanical properties of the biomaterial itself. Besides these efforts, techniques for sensing local oxygen level within the extracellular microenvironment have been developed to elucidate roles of major metabolites in the biological and physiological context. This study presents a strategy to control the distribution of soluble chemicals and to monitor oxygen concentrations within the bulk of a three-dimensional tissue scaffold. The presentation of this strategy includes: 1) a lithographic technique to build functional microfluidic structures directly within a calcium alginate hydrogel seeded with cells, 2) synthesis of oxygen sensing microbeads, 3) characterization of the lithographic process with respect to microstructural fidelity and cell viability, 4) characterization of convective and diffusive mass transfer of small and large solutes, and oxygen within this microfluidic scaffold, and 5) demonstration of temporal and spatial control of the distribution of non-reactive solutes and reactive solutes within the bulk of the scaffold. In vitro culture of primary chondrocytes and HepG2/C3A cell lines will be presented to illustrate the use of this new culture format to control and monitor cellular function.

Microfluidic cell fusion under continuous direct current voltage

Jun Wang and Chang Lu[a)]
Department of Agricultural and Biological Engineering, School of Chemical Engineering, Birck Nanotechnology Center, Bindley Bioscience Center, Purdue University, West Lafayette

Introduction

Cell fusion is a powerful tool for analysis of gene expression, chromosomal mapping, antibody production, cloning mammals and cancer immunotherapy. Cell fusion has been carried out based on several approaches. Chemical and virus-mediated fusion methods were developed in the earlier years based on the application of chemical fusogens such as polyethylene glycol or fusogenic virus (i.e. Sendai)[1, 2]. These methods have been associated with limitations such as toxicity to cells, batch-to-batch variability, and low efficiency. In comparison, electrofusion, which is based on the application of high voltage electric pulses, can be applied to a wide range of cell types with high efficiency and high post-fusion viability[3, 4]. Electrofusion typically requires specialized equipment which generates both low-voltage alternating current (AC) for cell alignment/contact and high-voltage direct current (DC) pulses for cell fusion[5]. Due to the complexity and cost associated with the instrumentation, very few studies have explored realizing this important technique on a microfluidic platform[6, 7]. All these demonstrations involved the application of electrical pulses.

Here, we demonstrate a technique which produces cell-to-cell electrofusion using a common DC power supply on a microfluidic platform. In our method, the cells were first conjugated based on biotin-streptavidin interaction. The electrofusion was then conducted by flowing the linked cells through a microfluidic channel with geometric variation under continuous DC voltage. We were able to achieve fusion efficiency comparable to that of conventional specialized equipment based on AC alignment and electrical pulses. The processing was carried out at single cell pair level.

Experiment

1. Microfabrication and General Procedures

The general information about PDMS microfluidic chip fabrication, culture of CHO-K1 cells, and the application of phase contrast and fluorescence microscopy was mostly provided in our previous publication[8]. The excitation and emission from cells labeled with calcein AM or SYTOX (Molecular Probes, Eugene, OR) were filtered by a fluorescence filter cube (exciter HQ480/40, emitter HQ535/50, and beamsplitter Q505lp, Chroma technology, Rockingham, VT). The excitation and emission from Hoechst 33342 (Molecular Probes, Eugene, OR) labeling were filtered by another filter cube (exciter D350/50, emitter D460/50, and beamsplitter 400dclp, Chroma technology, Rockingham, VT).

2. Modeling of the Electrical Field in the Device

We applied the Conductive Media DC model from Comsol 3.2 (COMSOL, Inc., Burlington, MA) to model the electric field distribution in Figure 1(b). Assuming no ion concentration gradient in the flowing fluid carrying the current, Ohm's law was used for current density calibration,

$$\nabla \cdot (-\sigma \nabla V) = 0 \qquad (1)$$

[a)]Author to whom correspondence should be addressed; email: changlu@purdue.edu

Where σ is the conductivity ($S\ m^{-1}$), V is the voltage. For our buffer system we used 1 S/m as the value of σ. We selected "electric potential" option as the boundary condition for the inlet and outlet in the software. All the walls were considered as electrically insulated.

3. Preparation of Cell Conjugates Based on Chemical Linking

Cells were harvested by scraping. It needs to be noted that we did not apply the common method of detaching cells using trypsin because cells treated with trypsin would have low affinity to Sulfo-NHS-LC-biotin.

The procedure of conjugating cells was similar to what was described in previous literature.[9] The cells were first washed by ice-cold PBS buffer (10mM phosphate buffer, 137 mM NaCl, pH adjusted to 8.0) twice to remove amine-containing culture medium and cell debris in the solution and then suspended in the same PBS (pH=8.0) buffer at a concentration of 5×10^7 cells/ml. The cells were then biotinylated by adding Sulfo-NHS-LC-biotin (Pierce, Rockford, IL) to a final concentration of 50 μg/10^6 cells. The cells were incubated at room temperature for 30 mins with occasional gentle shaking to prevent cells from aggregation. After biotinylation, cells were resuspended in the PBS buffer (pH=7.4) with 100 mM glycine added for quenching unreacted Sulfo-NHS-LC-biotin residues. One half of the cell sample was transferred to 4 °C water bath for future cell conjugation. The other half of the cell sample was washed by PBS buffer (pH = 7.4) for 2 times and then treated for streptavidin coating. Streptavidin in 5 mg/ml stock solution was added to the sample to a concentration of 1 mg/10^7 cells. The cells were incubated at room temperature for 25 mins with gentle shaking. The two cell samples (one coated with biotin and the other coated with biotin-streptavidin) were washed twice and resuspended in the electrofusion buffer (1 mM $MgSO_4$, 8 mM Na_2HPO_4, 2 mM KH_2PO_4, and 250mM sucrose, pH=7.2) at 5×10^7 cells/ml before being mixed for cross-linking. The mixed cells were gently concentrated at 300g for 2-5 seconds until a fraction of the cells precipitated to the bottom of the tube. The sample was then incubated for 15 mins. The cell sample was carefully diluted by the electrofusion buffer to 10^7 cells/ml before the electrofusion experiment.

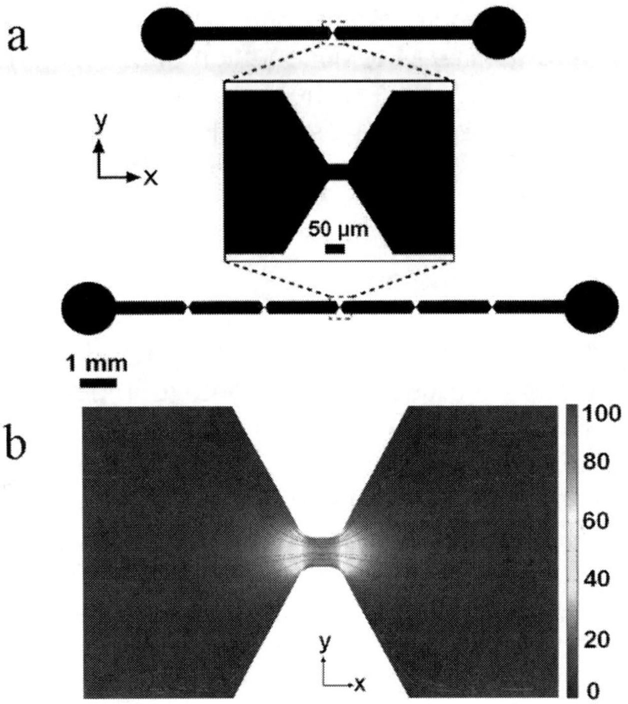

Figure 1 (**a**) The layout of single-pulsed and five-pulsed microfluidic devices for cell electrofusion. As shown in the inset figure, the narrow section was 50 μm long and 40 μm wide. The wide sections had a width of 400 μm. In both devices, the depth of the channel was uniformly 33 μm. The single-pulsed channel was 8.2 mm long and the five-pulsed device was 13.2 mm long. (**b**) The modeling of the electrical field intensity in a microfluidic structure with alternating wide and narrow sections (Comsol 3.2). The dimensions of the narrow section and the width of the wide sections were the same as those in (**a**) and the total length of the structure was 800 μm. We arbitrarily assigned 100 to the highest field strength in the structure.

Typically 50-55% of the cell population was conjugated after these steps, with more than half of them being one-to-one conjugation.

To facilitate the observation of cell fusion, in some experiments (as shown in Figure 2(a) and Figure S2) half of the cells involved in the electrofusion (either biotin coated or biotin-streptavidin coated) were labeled by a fluorogenic dye, calcein AM (Molecular Probes, Eugene, OR). The labeling was done by incubating the cells with calcein AM at a concentration of 1 μg/ml for 10 mins.

4. Cell Electrofusion

Prior to the experiment, the microfluidic channel was flushed with the electrofusion buffer (1 mM $MgSO_4$, 10 mM phosphate buffer, and 250mM sucrose, pH=7.2) for 15 minutes to condition the channel and remove impurities. The inlet of the channel was connected to a syringe pump (PHD infusion pump, Harvard Apparatus, Holliston, MA) through plastic tubing. The pump rate was in the range of 45~225 μl/hr. If we only consider the contribution to the cell velocity from the flow rate of the buffer, the durations for cells to be in the narrow section would be 5.3, 2.6, and 1.0 ms when the flow rates are 45, 90, and 225 μl/hr, respectively. However the actual durations (pulse widths) were shorter than the above numbers and varying with the field intensity, due to the contribution to the cell velocity from the electric field. A high voltage power supply (PS350, Stanford Research Systems, Sunnyvale, CA) was used to generate a direct current (DC) electric field inside the channel. The duration of the electrofusion experiment was 1~3 mins until the receiving reservoir contained enough cells for further analysis. Longer processing time may cause significant change in the buffer pH.

5. Fusion Index

Cells were stained by incubation with Hoechst 33342 (1 μg/ml) for 5 mins before electrofusion. Cells were transferred to a 96 well plate immediately after electrofusion and observed (within 1 hr after the electrofusion) under an inverted fluorescence microscope (objective 40×). The number of nuclei per cell and the number of cells containing *n* nuclei were counted. Usually about 500 to 1000 cells were enumerated for the calculation of fusion index in one trial and two trials were conducted for one data point in Figure 3(a).

6. Cell Viability

We determined cell viability after electrofusion using SYTOX exclusion. SYTOX (Molecular Probes, Eugene, OR) is a membrane-impermeant nuclear stain excluded by living cells. Cells were collected from the receiving reservoir (the outlet) immediately after electrofusion and transferred to a 96 well plate with PBS buffer (pH=7.4). The cells were incubated in the PBS buffer with 1 μM SYTOX added for 10 mins before the viability was determined (1 hr after electrofusion). Usually about 500 to 1000 cells were enumerated for the calculation of percentile viability in one trial and two trials were conducted for one data point in Figure 3(b). It needs to be noted that we measured only short-term cell death (1 hr after the electrofusion) and the long-term viability could be different.

Results and Discussion

As shown in Fig.1(a), an electrofusion device in our design consisted of a microfluidic channel with narrow and wide sections. Devices with one or five narrow sections were tested in this work. As we and other researchers previously showed, such channels generate local field intensities which are determined by the channel geometry[8, 10, 11]. Fig. 1(b) shows the

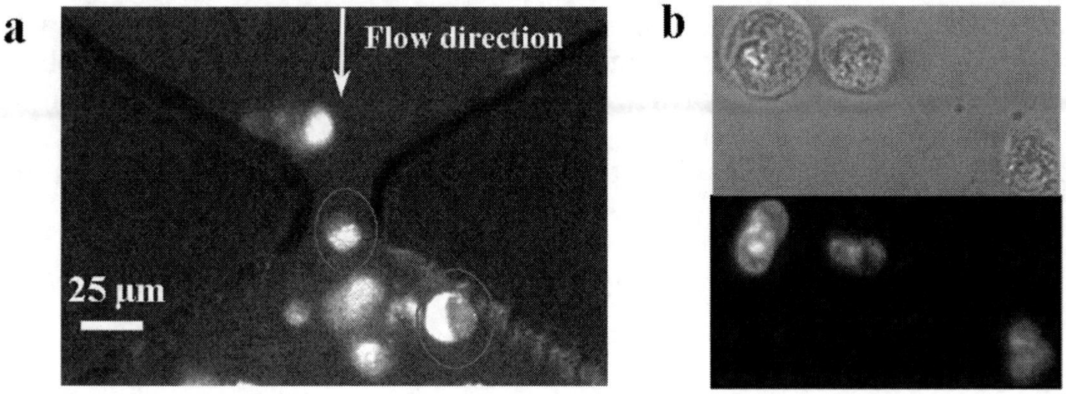

Figure 2 (**a**) A fluorescent image of fused cells (circled by red, between one cell labeled by calcein AM and one unlabeled) immediately after flowing through the narrow section. The narrow section here was 25 µm wide and 30 µm long (slightly different from the design in Figure 1) and the field intensity in the narrow section was 1200 V/cm. (**b**) Fused cells labeled by Hoechst 33342. The upper picture is the phase contrast image and the lower picture is the fluorescent image of the same cells. They were generated in a single-pulsed device with 700 V/cm in narrow section and a flow rate of 90 µl/hr.

modeling of the field intensity inside a microfluidic channel with one narrow section when a DC voltage is established across the channel. The modeling suggests that the field strength at the center of the narrow section is around 9.7 times higher than the field strength in the bulk of the wide sections (200 µm or further away from the narrow section). This number is roughly the ratio between the width in the wide section(s) and the one in the narrow section. In principle we can control the overall voltage so that only the field in the narrow section(s) is high enough for cell fusion and the field in the rest of the channel is too weak to have adverse effects on the viability of cells. When cells flow through the device, they experience field intensity variations equivalent to electrical pulse(s). The "pulse width" is determined by the length of the narrow section and the velocity of cells. The microfluidic chips were fabricated based on polydimethylsiloxane (PDMS) using standard soft lithography method[12]. We applied Chinese hamster ovary (CHO-K1) cells in our experiments. We demonstrated cell fusion between CHO cells. In principle the same procedure and device can be applied to fuse two different cell types and obtain hybrid cells as the result. The cells were first chemically linked based on the interaction between biotin and streptavidin using a protocol similar to what was reported in the literature [9, 13]. Briefly, the plasma membrane of cells was biotinylated first. One half of the cell sample was then coated with streptavidin before it was mixed and incubated with the other half of the cell sample. After this step, about 50-55% of the cell population was conjugated. The linked cells had intact membrane between them and there was no exchange of the intracellular contents between different cells. The linked cells were suspended in electrofusion buffer (1 mM $MgSO_4$, 10 mM phosphate buffer, and 250mM sucrose, pH=7.2), then flowed into the electrofusion devices (single-pulsed or five-pulsed) under the pressure from a syringe pump, while a continuous DC voltage was established across the channel.

Two approaches were used to observe the cell fusion. First, we labeled half of the cells with a fluorogenic dye, calcein AM (Molecular Probes, Eugene, OR, following the manufacturer's protocol) and left the other half unlabeled before the chemical conjugation. Cell fusion between labeled cells and unlabeled cells was observed immediately after they flowed through the narrow section [Fig. 2(a)]. Calcein (the fluorescent derivative of calcein AM) was observed to diffuse into the other half of the fused cell within minutes. In the second approach, we stained the cell nuclei using a nuclear counterstain, Hoechst 33342 (Molecular Probes, Eugene, OR) and observed the number of nuclei in cells after electrofusion [Fig. 2(b)]. A fraction of the cells contained two or more nuclei. The efficiency of cell fusion is characterized using fusion index (FI) which is defined as the fraction of nuclei in polynucleated cells in the total number of nuclei and is calculated using equation (1) below

$$FI(\%) = \frac{\sum_{n=2}^{\infty} nC_n}{\sum_{n=1}^{\infty} nC_n} \times 100 \qquad (2)$$

where C_n is the number of cells containing n nuclei[14]. We observed 2 or 3 nuclei in the vast majority of the fused cells. It needs to be noted that a fraction of polynucleated cells is due to cell division. Our previous studies showed that the field in the wide sections, which was substantially lower than the threshold for electric breakdown of the membrane, did not affect the cell viability significantly[8]. We varied the electrofusion field in the narrow section(s) in the experiment. The duration of exposure or the "pulse width" in the narrow section(s) was also varied by changing the flow rate controlled by the syringe pump. If we only consider the contribution to the cell velocity from the flow rate of the buffer, the durations for cells to be in the narrow section would be 5.3, 2.6, and 1.0 ms when the flow rates are 45, 90, and 225 μl/hr, respectively. However the actual durations (pulse widths) were shorter than the above numbers and varying with the field intensity, due to the contribution to the cell velocity from the electric field. Due to the difficulty in determining the cell velocity in the narrow section(s) experimentally, we specified the "pulse widths" by indicating the pumping rates [Fig. 3]. The fusion index among viable cells was calculated [Fig. 3(a)]. The fusion index was around 10-15% when there was no electric field due to the division in the cell population[15]. In a single-pulsed device, the fusion index increased remarkably when the field strength in the narrow section became higher during the processing. When the field intensity was increased to 1200 V/cm, the fusion index was up to 44% (around 30% after deducting the fraction due to cell division) at all three flow rates. The pulse width made a significant difference when the field intensity was between 600 and 1000 V/cm. The longer pulse width (at lower flow rate) resulted in higher fusion efficiency. We also carried out the cell fusion in the five-pulsed device. The

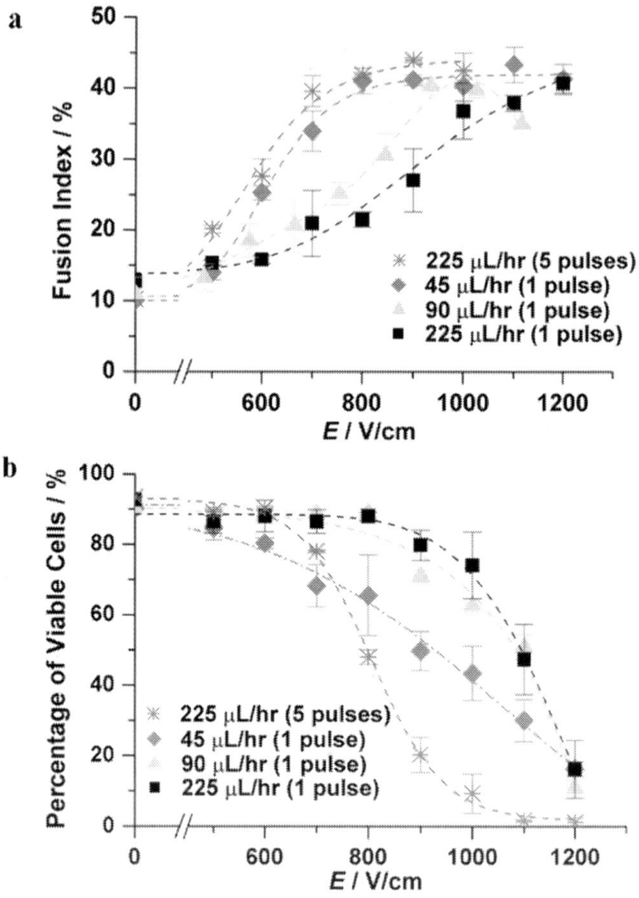

Figure 3 (**a**) The fusion index (among viable cells) at different electrofusion field strengths and flow rates in the single-pulsed and five-pulsed devices. (**b**) The percentage of viable cells measured under the same conditions as in (**a**). Trend lines are added to guide the eye.

application of multiple pulses improved the efficiency of cell fusion significantly. The five-pulsed device yielded fusion indexes that were consistently higher that those resulted from a single pulse of the same pulse width. The efficiency of cell fusion by our approach was comparable to the result obtained using conventional pulse generator on the same cell type and similar buffer system[15]. We also examined the viability of cells (both fused and non-fused cells) under these conditions using SYTOX green exclusion[8] [Fig. 3(b)]. The viability of cells in general decreased with increasing field strength and pulse width. The use of five-pulsed device created a marked decrease in the cell viability. It needs to be noted that the optimum fusion parameters have to be empirically derived for each cell system because they vary considerably from species to species [16].

Conclusion

We demonstrated that a simple microfluidic device could effectively produce cell fusion under continuous DC voltage, when coupled with chemical linking method. This technique offers efficiency comparable to that of conventional electrofusion technique which requires expensive and complicated apparatus. Cell fusion was conducted at single pair level in the microfluidic channel and the physical dimensions of the channel effectively prevented more than 3 cells from fusing together. These features are important when cell fusion needs to be conducted based on scarce cell sources such as primary cells. This demonstration will help establish microfluidics as a viable platform for studying a number of important biological problems involving cell fusion. Further studies are needed to characterize the properties of fused cells produced by this technique.

Acknowledgement

This research was supported by a Purdue Research Foundation fellowship and Purdue University.

References

1 White, J., Matlin, K. and Helenius, A. (1981) Cell fusion by Semliki Forest, influenza, and vesicular stomatitis viruses. *Journal of Cell Biology,* **89**, 674-679

2 Davidson, R. L., O'Malley, K. A. and Wheeler, T. B. (1976) Polyethylene glycol-induced mammalian cell hybridization: effect of polyethylene glycol molecular weight and concentration. *Somatic Cell Genetics,* **2**, 271-280

3 Neumann, E., Gerisch, G. and Opatz, K. (1980) Cell-fusion induced by high electric impulses applied to *Dictyostelium. NATURWISSENSCHAFTEN,* **67**, 414-415

4 Zimmermann, U. and Vienken, J. (1982) Electric field-induced cell-to-cell fusion. *Journal of Membrane Biology,* **67**, 165-182

5 White, K. L. (1995) Electrofusion of mammalian cells. In Methods in Molecular Biology (Nickoloff, J. A., ed.), pp. 283-294, Humana Press Inc., Totawa, NJ

6 Tresset, G. and Takeuchi, S. (2004) A microfluidic device for electrofusion of biological vesicles. *Biomedical Microdevices,* **6**, 213-218

7 Stromberg, A., Karlsson, A., Ryttsen, F., Davidson, M., Chiu, D. T. and Orwar, O. (2001) Microfluidic device for combinatorial fusion of liposomes and cells. *Analytical Chemistry,* **73**, 126-130

8 Wang, H. Y. and Lu, C. (2006) Electroporation of mammalian cells in a microfluidic channel with geometric variation. *Analytical Chemistry,* **78:** 5158-5164

9 Bakker Schut, T. C., Kraan, Y. M., Barlag, W., de Leij, L., de Grooth, B. G. and Greve, J. (1993) Selective electrofusion of conjugated cells in flow. *Biophysical Journal,* **65**, 568-572

10 Jacobson, S. C., Culbertson, C. T., Daler, J. E. and Ramsey, J. M. (1998) Microchip structures for submillisecond electrophoresis. *Analytical Chemistry,* **70**, 3476-3480

11 Plenert, M. L. and Shear, J. B. (2003) Microsecond electrophoresis. *Proceedings of the National Academy of Sciences of the United States of America*, **100**, 3853-3857

12 Duffy, D., McDonald, J., Schueller, O. and Whitesides, G. (1998) Rapid prototyping of microfluidic systems in poly(dimethylsiloxane). *Analytical Chemistry,* **70**, 4974-4984

13 Tomita, M. and Tsong, T. Y. (1990) Selective production of hybridoma cells: antigenic-based pre-selection of B lymphocytes for electrofusion with myeloma cells. *Biochimica et Biophysica Acta,* **1055**, 199-206

14 Poste, G. and Papahadjopoulos, D. (1976) Fusion of mammalian cells by lipid vesicles. *Methods in Cell Biology,* **14**, 23-32

15 Ramos, C., Bonenfant, D. and Teissie, J. (2002) Cell hybridization by electrofusion on filters. *Analytical Biochemistry*, **302**, 213-219

396d Converging Flow Electroporation For Drug And Gene Delivery

Shengnian Wang, Xulang Zhang, Chee Guan Koh, and J. James Lee. Department of Chemical and Biomolecular Engineering, The Ohio State University, 140 West 19th Avenue, Columbus, OH 43210

Electroporation is an efficient tool to introduce nucleic acids into cells or tissues. Together with microfluidic technology, it offers several advantages, such as low but uniform electric parameter, in situ analysis. For clinical applications, there is a need to transfect a large number of cells (e.g. >109 cells). This is difficult to achieve using the current bulk electroporation device where the recommended cell population is around 105–106. A continuous electroporation device that could provide high cell throughput with a mild and uniform electro pulse would be highly desirable. In this presentation, a continuous converging microfluidic process is established for electroporation with high transfection efficiency and cell viability. When cells sequentially pass through the throat region of the converging microchannels, they will experience acceleration and a high electric field there, similar to a single electric pulse in the bulk electroporation process. Because of its continuous operation and the confinement of the microchannels, every cell experiences the same level of electroporation and, consequently, the delivery of therapeutic materials can be accomplished uniformly for a large number of cells. To gain the advantages of electric field localization and cell/gene confinement, cells and therapeutic materials are encapsulated in nanoporous microcapsules. Essentially, this is a three-dimensional analogy to a two-dimensional membrane-(cell/therapeutic materials)-membrane sandwich structure. It also shows that microencapsulated cells grow better than conventional methods, because of the in vivo-like microenvironment provided by the microcapsules.

396e Design Considerations for Dynamic Field Gradient Focusing

Jeffrey M. Burke and Cornelius F. Ivory. Chemical Engineering, Washington State University, Pullman, WA 99163

Dynamic Field Gradient Focusing (DFGF) is an equilibrium gradient method that utilizes an electric field gradient to simultaneously separate and concentrate charged analytes Unlike isoelectric focusing which separates proteins in the presence of a pH gradient, DFGF does not require that the analytes be amphoteric, only that they have differences in their individual electrophoretic mobilites. The electric field gradient is generated by a computer-controlled electrode array that allows for dynamic control and manipulation of the field profile during the course of an experiment to increase peak resolution, migrate analytes to off-take ports or to systematically elute individual species. Over the past few years interest in focusing methods, such as DFGF, has soared. However, wide acceptance of these techniques has been slow. This is due in most part to several bottlenecks that have led to lower than expected resolution and peak capacity. Issues which cause distortions in the electric field profile or lead to smaller than anticipated fields are of specific interest because they can have the greatest impact on the separation performance of DFGF. Experimental results of peak profiles suggest a sudden loss of peak resolution during elution. Corresponding electric field measurements indicate that changes in the field profile near the cathodic end of the separation channel could be responsible. This paper discusses these problems and then uses Comsol Multiphysics simulations and experimental validation to show how the DFGF system can be redesigned to eliminate field distortions that lead to decreases in system performance.

396f Effect Of Valence On Optimal Separation Times Of Biomacromolecules For An Electrokinetic Couette-Based Separator

Jennifer Anne Pascal, Chemical Engineering, Tennessee Tech University, PH-214, Cookeville, TN 38505, Mario A. Oyanader, Universidad Catolica del Norte, Avenida Angamos, Antofagasta, 0610, Chile, and Pedro E. Arce, Chemical Engineering, Tennessee Technological University, Department of Chemical Engineering, P.O. Box 5013, Cookeville, TN 38505.

Separation of biomacromolecules for medical diagnostics, pharmaceutical applications, and bioprocesses is a multi-million dollar effort where even the minutest increase in the separation efficiency has a very healthy economical impact on the process or application. Antibiotics, proteins, and DNA are some key examples of biomacromolecules where new techniques and approaches for separation and or characterization are required or needed.

In this presentation, the authors will summarize recent work on models based on fundamental principles that allow for a systematic understanding of the role of valence in improving the optimal separation time of biomacromolecules. A scaling-based argument, rooted in area-averaging approaches, and convective-diffusive transport will be used to compute optimal separation times. These results will predict the effects of valence on these separation times and may be used to design and or optimize a Couette-based separation device. We will illustrate the approach for idealized geometries of capillary channels with proteins as model molecules.

Studies of electroosmotic mobilities and zeta potentials in plasma polymerised microchannel surfaces

Malinda Salim, Brian O'Sullivan, Gautam Mishra, Gregory J.S. Fowler, Phillip C.Wright and Sally L.McArthur

Microfluidic systems are becoming increasingly important for bioanalytical and biochemical research such as proteomics, genomics, clinical diagnostics and drug discovery. Its advantages over macroscale systems include fast analysis time and lower sample consumption. These microfluidic devices usually require the use of electroosmotic flow (EOF) as the mean of fluid transport, although variations and inconsistencies in EOF may interfere with device performances. The stability of EOF in microfluidic devices is usually poor and mainly affected by the adsorption of buffer as well as sample components onto the surfaces. Excessive rinsing and etching with solutions such as NaOH and HCl is usually required to obtain reproducible EOF in repetitive runs. Control of EOF is therefore crucial where one usually resort in surface modification. In addition to this, control of the non-specific adsorption of biomolecules onto device surfaces through surface modification is also an important issue since any adsorbed molecules will cause non-uniform surface charge and thus EOF, depending on the nature and extent of adsorption. Physical adsorption and covalent grafting are usually employed as surface modification strategies. Both techniques have their unique advantages and disadvantages. Physical adsorption for example, despite its easier applications, often suffers from the lack of stability and may alter system selectivity. Although covalently grafted surfaces are more advantageous in terms of stabilities compared to physically adsorbed ones, they often require complicated procedures and are time-consuming. Difficulties in surface modification often arise due to the large variations in chip construction materials ranging from glass, fused silica, quartz, silicon, to lack of functional group polymeric materials.

Microchannel surface modification by plasma polymerisation method may also be an attractive alternative approach. Plasma polymerisation can be used to deposit thin films of various structures and functionalities. It offers several advantages over other techniques in its ability to modify almost any substrates without altering the bulk properties, has a large selection of monomers, solvent-free and a relatively simple one-step coating procedure. Plasma polymerisation is a relatively complex process where repetitive fragmentation, rearrangement, crosslinking, ionisation & polymerisation of the monomers. Plasma polymerisation of tetraethylene glycol dimethyl ether (tetraglyme) onto glass microfluidic channel and PTFE base has been shown successful, which has polyethylene oxide (PEO)-like non-fouling characteristics. X-ray photoelectron spectroscopy (XPS), direct enzyme-linked immunosorbent assay (ELISA) and fluorescent imaging were used to investigate the channel coatings. Successful plasma deposition of tetraglyme onto microchannel surfaces were shown with very low or almost no fibrinogen adsorbed (0-0.1% Nitrogen). Water contact angle measurements and XPS studies revealed good stability of tetraglyme coating upon rinsing and shelf-storage of more than 3 months. This plasma polymerised tetraglyme surface is hydrophilic having an advancing water contact angle of ~ 50^0 as measured using water contact angle goniometer of 2 μl droplet.

Our study will now extend to investigate the electroosmotic behaviour and zeta potentials of this plasma polymerised tetraglyme (ppTG) surface under the effects of various surrounding buffer solution pH from 3-10. Streaming potential measurements on flat surfaces were used to corroborate the zeta potential values obtained from the EOF measurements using Smoluchowski equation. Other types of plasma polymerised surfaces, i.e. cationic (plasma polymerised allylamine, ppAAm) and anionic (plasma polymerised acrylic acid, ppAAc) were also investigated. Comparison examples were then carried out between the plasma polymerisation technique (cationic ppAAm surface) and the conventionally used physical adsorption and covalent graft method, i.e. the physically adsorbed cationic polymer polyethyleneimine (PEI), and covalent grafting of PEI onto pp.AAc surfaces using EDC/NHS chemistry respectively. Results indicate EOF suppression compared to uncoated microchannel surfaces, with isoelectric points of tetraglyme, acrylic acid and allylamine at ~ pH3.8, 3.5 and 8.9 respectively. The surface stabilities of pp.AAm are comparable to covalently grafted amine containing polymers (> 3 days) while physically adsorbed polymers show shorter stabilities (< 2 days). Other plasma polymerised surfaces also exhibit high stabilities enabling EOF runs of more than 3 days without deterioration.The overall purpose of these studies serves to investigate the flow and surface properties of several plasma polymerised surfaces, and their viability towards EOF device applications.

458b Size-Based Separation Of Proteins By Microchip Electrophoresis Using Acid-Labile Surfactant As A Replacement For SDS To Aid Ce-MS Integration

Brian E. Root, Materials Science and Engineering, Northwestern University, 2145 Ridge Ave, E136, Evanston, IL 60208 and Annelise E. Barron, Chemical and Biological Engineering, Northwestern University, 2145 Sheridan Road, Evanston, IL 60208.

The realization of an integrated lab-on-a-chip system for the analysis of proteins has been hindered by several non-trivial challenges. Among these challenges are that the reagents required for one process can often be incompatible with a downstream process. One highly researched technique is the integration of electrophoretic separations of proteins and analysis of the proteins by mass spectrometry. Sodium dodecylsulfate (SDS) is often needed to solubilize hydrophobic proteins and provides a constant charge-to-mass ratio resulting in a purely size-based separation. However, a drawback to this system is that SDS has been shown to interfere with mass spectrometry. In this study, an acid-labile surfactant (ALS) was investigated as a replacement for SDS in the electrophoretic separation of proteins on a microfluidic chip. ALS is advantageous as it degrades at low pH (pH < 3), and therefore may facilitate the integration of electrophoresis with mass spectrometry. We show the separation of proteins between 18 kDa and 116 kDa on a glass microfluidic chip dynamically coated with poly-*N*-hydroxyethylacrylamide. The proteins were labeled with the Py-1 dye, which does not alter the charge of the protein, and detected using laser-induced fluorescence. Multiple field strengths and SDS or ALS concentrations were tested to determine their effects on protein separation. Separation efficiency and migration times of a protein ladder are compared using SDS and ALS as the surfactants in the running buffer. While SDS provides a higher separation efficiency, the separation achieved by ALS appears to be sufficient to be used as a sized-based separation method prior to mass spectrometry. The separation efficiencies for proteins with ALS as the surfactant maintained peak efficiencies between 10^4 and 10^5 plates / meter. This demonstration of an acid-labile surfactant as a replacement for SDS in the size-based separation of protein will aid the integration of microchip electrophoresis with mass spectrometry.

458c Differential Release of Intracellular Molecules during Electrical Lysis Observed at the Single Cell Level

Ning Bao[1], Jun Wang[1], and Chang Lu[2]. (1) Agricultural and Biological Engineering, Purdue University, West Lafayette, IN 47907, (2) Agriculture and Biological Engineering, Chemical Engineering, Purdue University, West Lafayette, IN 47906

Analysis of intracellular materials at the single cell level presents opportunities for probing the heterogeneity of a cell population. Electrical lysis has been investigated as a rapid method for disruption of the cell membrane in the literature. In a typical single cell analysis experiment, electrical lysis provides the means for releasing the intracellular materials which are subsequently analyzed by assays such as electrophoresis. In this report, we studied the release of intracellular materials at the single cell level during electrical lysis. We examined the release of a small molecule, calcein (MW~600), and a 72 kDa protein kinase, Syk, tagged by enhanced green fluorescent protein (EGFP) from chicken B cells during electrical lysis at the single cell level. We studied the effects of the field intensity and field duration on the release of the two molecules. We found that calcein in general was released at lower field intensities and shorter durations than did SykEGFP. The field duration had more effect on the release of calcein than on that of SykEGFP. Taking advantage of the phenomena, we were able to deplete calcein from the cells before SykEGFP started to be release. Our results indicate that the release process during electrical lysis can vary significantly from one molecule to another. This needs to be taken into consideration when electrical lysis is performed for single cell analysis.

458d A Fractal Analysis of the Binding Kinetics of the Heat-Shock Protein Chaperone Dnak on a Spr Biosensor Surface

A. Sadana[1], Reema Taneja[2], and Kennon C. Shelton[2]. (1) University of Mississippi, Rebel Drive, University, MS 38655-1848, (2) Chemical Engineering Department, University of Mississippi, Rebel Drive, University, MS 38677-1848

A fractal analysis is presented for the binding kinetics of the Heat-Shock Protein Chaperone DnaK in unstressed and stressed Escherichia coli cell lysate to anti-DnaK on a surface plasmon resonance (SPR) biosensor chip surface or by ELISA (Vostiar et al., 2003). The binding kinetics is described by either a single-, dual-, or a triple-fractal analysis. As expected, the levels of chaperone DnaK expressed in stressed Escherichia coli cell lysate is higher than that expresed in unstressed Escherichia coil cel lysate. There is a change in the binding mechanism as one goes from a single- to a dual- to a triple-fractal analysis. The analysis provides values of the binding rate coefficient, k and the degree of heterogeneity on the sensing surface. Furthermore, the analysis provides fresh physical insights into the binding of Escherichia coli chaperone DnaK to anti-DnaK under stressed and heat-stressed conditions. It is of interest to note that for the unstressed Escherichia coli cell culture analysis of DnaK by ELISA, there is a significant amount of variation in the DnaK concentration with time. It is this dilute nature of the DnaK concentration in solution that requires a higher degree of fractal analysis (dual- or triple-) to adequately model the binding kinetics.

458e Predicting Electrohydrodynamic Flow Rates In Capillaries: Effect Of Geometry And Electrostatic Potential

Jennifer Anne Pascal, Chemical Engineering, Tennessee Tech University, PH-214, Cookeville, TN 38505, Mario A. Oyanader, Universidad Catolica del Norte, Avenida Angamos, Antofagasta, 0610, Chile, and Pedro E. Arce, Chemical Engineering, Tennessee Technological University, Department of Chemical Engineering, P.O. Box 5013, Cookeville, TN 38505.

Electrokinetic-based methods are used in a variety of applications including drug delivery and separation of biomolecules, among others. Many of these applications feature a fibrous or a porous medium that can be modeled by using capillary bundle models to predict the behavior of the flow rates within the particular system. It is often assumed that, within these bundles, capillaries of idealized geometries exist, including rectangular, cylindrical, and annular. Predicting flows in capillaries may be accomplished by applying the principles of electrokinetics in conjunction with hydrodynamics. In this project, capillaries of rectangular and cylindrical geometries are examined, for which velocity profiles are determined and parametrically studied. Afterwards, the flow rates are computed. In order to gain a deeper understanding of how these predictions differ according to the geometries used, the two situations described below are studied.

First, a ratio of the side magnitudes of the cross-sectional area of the rectangular geometry is selected in order to yield analytical velocity profiles. Specifically, velocity profiles with a 1D spatial dependence are described. This allows for an effective comparison of the flow rates predicted by the two different geometries. In fact, the ratio of the electrokinetic contribution of the rectangular geometry to the cylindrical geometry is placed between two limits whose values depend on the inverse Debye length. For low values of this parameter, the pure hydrodynamic limit is reproduced while, for large value of the Debye length, a lower value is obtained.

Second, and in order to see how the limits identified for the 1D case (see above) are affected, 2D spatial-dependence velocity profiles are identified. This case is solved numerically for the velocity profiles and values of the flow rate are afterwards computed.

The presentation will describe details of the two cases identified above. The analysis will be illustrated with dimensionless functions of the volumetric flow rates for both geometries, and quantitative comparisons of the prediction of the two analyses will be discussed. These flow rates are useful to predict the optimal time for many applications, including drug delivery and separations and assess the potential error made by choosing a given geometry over another.

Nano-Particle Trap by Integrating AC Electroosmosis with Microcantilever

Nazmul Islam*
Department of Electrical and Computer Engineering
Northern Arizona University, AZ 86011, USA

Jie Wu
Department of Electrical and Computer Engineering
The University of Tennessee, Knoxville, TN 37996, USA

Abstract

Recently AC electroosmosis (ACEO) has received increasing interest to control and manipulate particles and fluids at the micro/nano-scale. ACEO can operate at relatively low voltages, which is suitable for integrated lab-on-a-chip systems. Particle trapping by ACEO has no dependence on particle properties, so nano particle trapping can be possible. On the other hand, microcantilevers are finding wide applications in detecting bio/nano-particles. However, their usage has been limited to highly concentrated samples to ensure sufficient deposition of agents onto cantilevers. This paper reports the integration of *in-situ* particle concentrators on microcantilevers. Only a thin metal layer on microcantilevers is required to generate microfluidic convection of particles from solution bulk onto microcantilever surfaces. A working prototype is presented in the paper. Preliminary experiments concentrating nano particles were conducted and the particle concentration effect has been experimentally verified using AFM probes as microcantilevers.

Keyword: AC electroosmosis, micro-cantilever, microfluidic, nano-particle trapping, atomic force microscope, resonance frequency.

*To whom correspondence should be addressed. Tel: (928)523-5262. E-mail: Nazmul.Islam@nau.edu

Introduction

AC electrokinetics is the study of particle and fluid motion induced by AC electric fields. In the last decade it has received increasing interest due to its important applications in micro total analysis systems and miniaturized biomedical devices. With the application of very low voltage (~1V_{rms}), AC electrokinetics can be utilized to control and manipulate particles and fluids at the micro/nano scale, which are very difficult to achieve with existing techniques, such as pressure driven flow.

Through the interaction of ac electric fields and particles/fluids, ac electrokinetics can be used to sort, separate and filter particles, and to pump and mix fluids. AC electrokinetics can operate at relatively low voltages, which is suitable for integrated lab-on-a-chip systems [1]. This paper investigates AC electroosmosis for manipulating nanofluids/ particles with an aim to provide a generic platform for the transport and concentration functions in microfluidic devices. It is envisioned that these concepts can be integrated with life science and biomedical technologies to develop a new generation of labs-on-a-chip, with high-efficiency particle manipulation and sensing, micropumps and microfluidic mixers.

Particle trapping is conducted adjacent to an electrolyte/electrode interface of the gold (Au) electrode pair. We have developed the capability to integrate AC electroosmotic trap and micro-cantilever (MC) in a microfluidic system. As ACEO concentrator has no dependence on particle properties, the method is expected to be applicable to nano-particles collection. The sensitivity as well as the detection time could be improved by orders of magnitude if a concentration trap could be embedded with cantilevers. For this reason we have integrated micro-cantilever with AC electroosmotic trap in a microfluidic chamber.

Nano Particle Trap by Microcantilever Integration

A nanometer layer of charges/ions is induced by an electric field at the interfaces of electrolytes and solids, whether the solid is a metal or a cell membrane. If there also exist electric fields parallel to the electrodes, the induced ions will migrate under the influences of the tangential fields, and produce osmotic microflows due to fluid viscosity [2]. The generated microflows convey nano particles from the bulk of the fluid onto the electrode surface. As shown in figure 1, the top and bottom electrodes are asymmetric, the tangential electric fields are generated which induces electro-osmotic fluid motion. For the first half cycle as in figure 1(a), the bottom electrode is at positive potential, and negative ions are induced at the interface of the electrode and fluid. These negative ions interact with the electric field and produce two counter-rotating vortices from electrode edges inward on the electrode surface, creating null point at the center of the electrode.

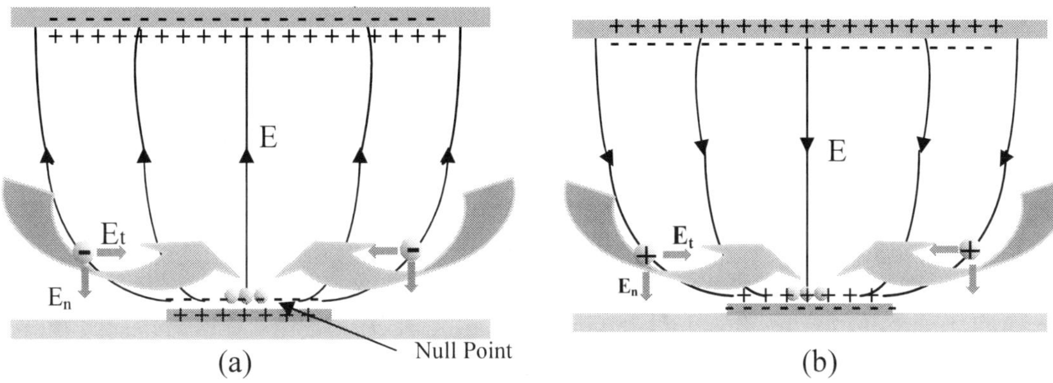

Figure 1: Concept of Parallel plate particle trapping for an AC cycle; (a) during the half cycle when the bottom electrode has positive polarity and ITO (indium-tin oxide) coated top electrode is negative; (b) next half cycle with opposite polarity. Flow motion and induced charges is also shown.

For the next half cycle as in figure 1(b), applied potentials switches polarity and the bottom electrode is at negative potential. Here the induced positive ions will interact with the electric field and again produces two counter-rotating vortices from electrode edges inward, and fluid motions are sustained thus the particles are trapped at the center of the electrode.

In our design of microcantilever trap, the metal-coated cantilevers substitute the patterned bottom electrode, so that particles will aggregate on the cantilevers [3, 4]. Figure 2 shows the experimental setup of cantilever particle trap. As shown in figure, photoresists (dielectrics) are coated on the conductive areas other than the cantilevers to suppress unwanted local EO flows. The ITO glass slide works as the top electrode, which is covering the whole fluid chamber. We have used Au-coated AFM probes as the MC, which has the dimension of 125μm X 30μm X 4μm. Tipped MC side was not used to avoid sharp electric fields.

The parallel plate design has been used to attract particles onto cantilevers for high sensitivity detection [4,5]. Because particle trapping/concentrating effect is more obvious with the smaller electrode [6], in our design the metal-coated cantilevers are facing one large electrode (covering a whole fluid chamber), so that particles will aggregate on the cantilevers.

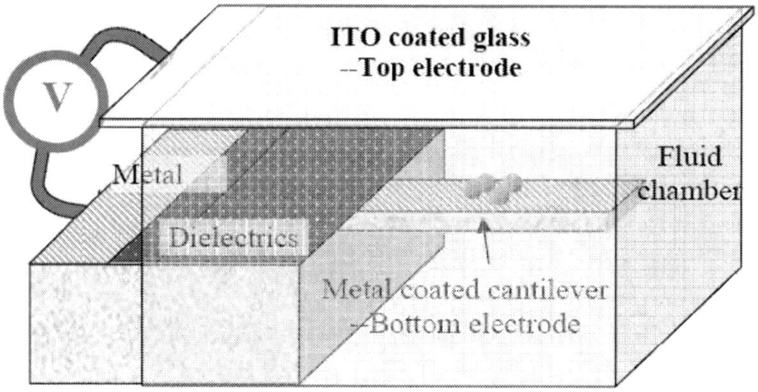

Figure 2: Experimental Setup of Cantilever particle trap

Figure 3 shows the experimental results of trapping 200nm fluorescent particles on MC. After applying the AC signal (100Hz, 400mVp-p), suspended particles accumulate at the center of the cantilever from all directions. As time passes, more fluorescent particles from the surrounding area accumulated and formed bright object pattern. After the particle trapping on the surface, the MC was dried with AC signals applied, so that particles will not get dispersed by diffusion, surface tension, etc. The particles were still concentrated at the center after fluid dryout. Then the particle trapping effect was verified with AFM resonance measurement.

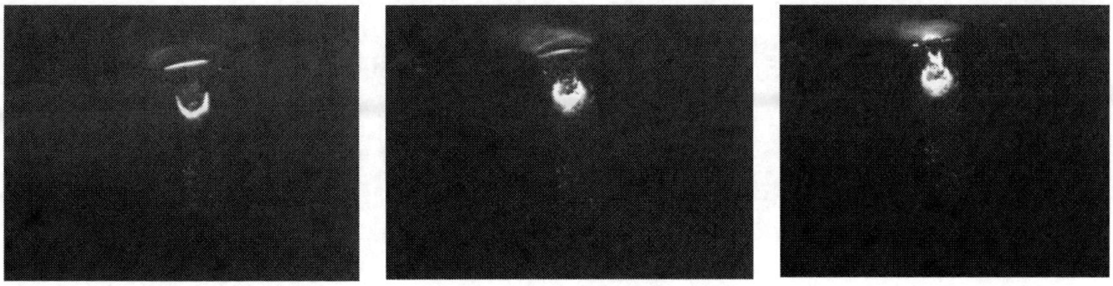

(a) **(b)** **(c)**

Figure 3: Image sequence of 200 nm fluorescent particles trapped on the micro-cantilever for AC Electroosmosis; (a) t = 10 min, (b) t = 20 min, (c) t = 30 min.

Result and Discussions

To verify the concentrated particle trapping on MC, we also measure the resonance frequency of MC before and after trapping experiments. MC resonance frequency is inversely proportional to the differential mass of cantilever [5]. The sensitivity of a cantilever to mass loading is mainly determined by the excited cantilever resonance frequency.

$$\Delta m = \frac{K}{4\pi^2}\left(\frac{1}{f'^2} - \frac{1}{f^2}\right) \tag{1}$$

Where, Δm is the mass change;
K=sensor spring constant;
f= resonance frequency before mass adsorption;
And f' is the resonance frequency during mass adsorption.

Changes in the mass and surface properties of the microcantilever through binding or hybridization of analytes to receptor molecules will directly influence its surface stress. This causes the microcantilever to deflect and the deflection is proportional to the analyte concentration and inversely proportional to mass loading. The more the particle concentration on ACEO-cantilever, the more is the bending. So the more mass on the cantilever means the lower resonance frequency. From our experimental result (Figure 4) we have got the resonance frequency of the MC goes down to 276.07 KHz after particle trapped on the MC for ACEO, which translates to a change of mass. For a change in mass we get,

$$\frac{f_1}{f_2} = \frac{\sqrt{m_{eff} + \Delta m}}{\sqrt{m_{eff}}} \tag{2}$$

Microcantilever dimension is 125μm x 30μm x 4μm. The volume of the MC is $1.5e^{-14}$ m^3. Microcantilever is Si based, and the density of Si is 2330 kg/m^3. So the mass of microcantilever is $3.495e^{-8}$ gm. By putting the frequency and mass values in equation (2), we have got an increase of 2.52% increase of mass for the frequency change from 279.52 KHz to 276.07 KHz. The change in resonant frequency as a function of the particle mass binding on the cantilever beam surface forms the basis of the particle detection scheme.

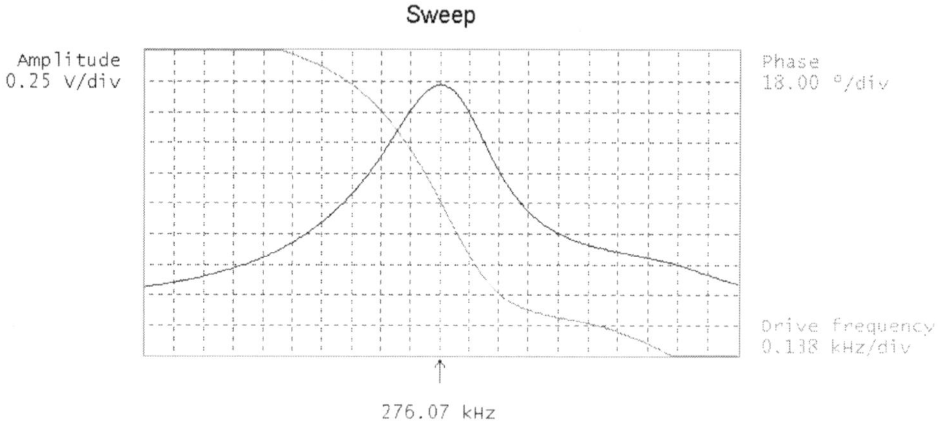

(a)

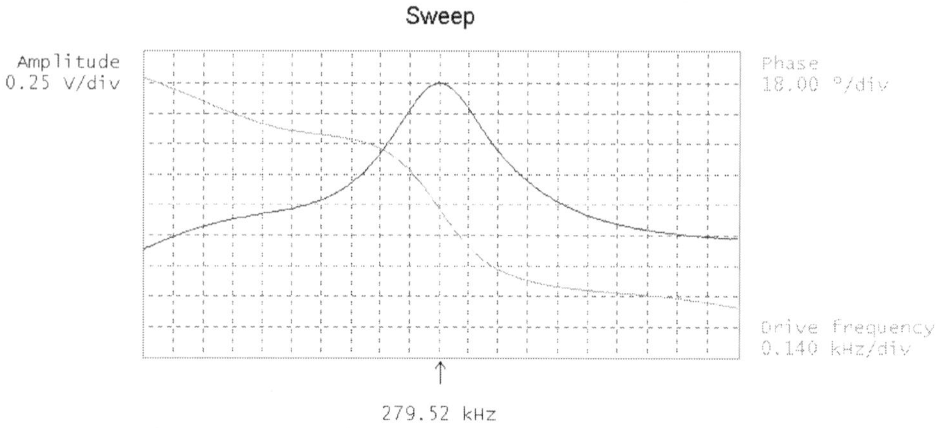

(b)

Figure 4. Resonance frequencies of the MCs (measured with multimode AFM.) (a) After the particle trapping by ACEO, 276.07 kHz, (b) Control experiment with no electric signal applied, 279.52 kHz.

Conclusions

A novel technique has been developed to trap nano particles. By combining both experimental investigation and theoretical analysis, this research work demonstrates a microcantilever particle trap. Interfacing the micro-cantilever with ACEO mechanism has expanded its capability for biological, physical and chemical detection and makes the whole system ultra sensitive. The research work substantially enriches the portfolio of transducers that can be used in high performance miniaturized analytical systems.

References

[1] A. Ramos, H. Morgan, N. G. Green and A. Castellanos, "AC electric-field-induced fluid flow in microelectrodes", J. Colloid Interface Sci. 217(1999) 420–2

[2] N. Islam, M. Lian, S. Swaminathan and J. Wu, "Micro/Nano- Particulate Fluid Manipulation in AC Electro-Kinetic Lab-on-a-Chip," 2nd ASM - IEEE EMBS Conf. Bio, Micro & Nanosyst., Jan. 15-18, 2006, San Francisco, CA, USA.

[3] J. Wu, N. Islam and M. Lian, "High Sensitivity Particle Detection By Biased AC Electro-Osmotic Trapping on Cantilever," 19th IEEE Int'l Conf. Micro Electro Mechanical Systems (MEMS 2006), Jan. 22-26, 2006, pp. 566 – 569, Istanbul, Turkey.

[4] N. Islam, M Lian, and J. Wu, "Enhancing Microcantilever Sensitivity with Integrated AC Electroosmotic Trapping," Microfluidics and Nanofluidics Journal, Volume 3, Number 3, June 2007.

[5] Sepaniak M, Datskos P, Lavrik N, Tipple C, "Microcantilever Transducers: A New Approach in Sensor Technology," Analytical Chemistry, November 1, pp. 568A-575A, 2002.

[6] Green N G, Ramos A, Gonzalez A, Morgan H, and Castellanos A, "Fluid flow induced by nonuniform ac electric fields in electrolytes on microelectrodes. III. Observation of streamlines and numerical simulation," Physical Review E 66, 026305, 2002.

521b Integrated Dielectrophoretic Chip For Continuous Sorting, Trapping, And Detecting Using Surface-Enhanced Raman Scattering

Diana S. Hou[1], I-Fang Cheng[2], Hsien-Chang Chang[2], and Hsueh-Chia Chang[1]. (1) Chemical and Biomolecular Engineering, University of Notre Dame, 182 Fitzpatrick Hall, Notre Dame, IN 46556, (2) Institute of Nanotechnology and Microsystem Engineering / Institute of Biomedical Engineering, National Cheng Kung University, Tainan, Taiwan

We report an integrated dielectrophoretic chip that can continuously sort, trap and detect pathogens with a throughput of 3 μL/min. A processing rate of 500 bacteria per second is estimated when 100 μl of a heterogeneous colony of 10^7 CFU(colony forming units)/ml is processed in a single pass within 30 minutes. Using planar electrodes to form three-dimensional dielectrophoretic (DEP) gates, the pathogens can be deflected to different channels on the chip where they are individually concentrated. Individual bacteria have different DEP characteristics based on their surface properties, permittivities, and whether they are alive or dead. Therefore, a mixture of bacteria can be sorted into individual channels using DEP based on these inherent characteristics. A DEP gate is also used to trap and concentrate the bacteria as they flow through the channel, forming a concentrated mound. Upon concentration Raman spectroscopy can detect and identify the concentrated mounds of bacteria. With the addition of silver nanoparticles into the pathogen mixtures, the Raman shift can be enhanced orders of magnitude through surface-enhance Raman scattering creating an rapid on-chip detection method. With electrodes on both the top and the bottom of the channel, the field can penetrate across the entire height of the channel. Consequently the DEP force will be exerted across the entire height of the channel in contrast to being restricted to the electrode surface as typically seen in parallel planar electrodes. These electrode gates can deflect and trap the pathogens as they pass through the channel when an AC field is applied at a frequency such that the particle exerts a negative DEP force. Thus the pathogens will be repelled from the high field region, which is the electrode itself, and be forced to cross fluid streamlines and deflect towards individual channels. However, there is a balance in the forces between the projected DEP force and the Stokes drag on the particle in the flow direction. A DEP gate perpendicular to the streamline requires a DEP force larger than the Stokes drag to prevent the particle from passing through the gate. At the high fields needed for this strong of a DEP force, reactions, heating or other damaging side effects to the electrodes and to the pathogens themselves can occur. However, an angled electrode will only need to exert a DEP force large enough deflect and slow the pathogen's velocity such that instead a sudden stop in the particle flow, the particle can translate along the electrode edge. Thus the pathogens can be sorted by these angled electrodes such that they are deflected to different channels on the chip. Concentration of the pathogens in the trapping electrode gate is also investigated using three different electrode designs, a crescent, a pointed tip, and a multi-curved "flower" shape. The angle of the electrode with respect to the streamline can be optimized for higher throughput of the pathogens without increasing the electric field to the point of damaging them or the integrity of the electrodes. Mixtures of *Escherichia coli Nissle*, *Lactobacillus*, and *Candida albican* are sorted, concentrated and detected using surface-enhanced Raman scattering. Differing spectra in each concentration will allow the rapid identification of the concentrated pathogen at the trapped gates. It is hence possible to expand the mixtures to include more than three types of bacteria, such as those found in a realistic gastrointestinal sample: *Escherichia coli Nissle*, *Lactobacillus plantarum*, *Enterobactor cloacae*, *Citrobacter rodentium*, *Proteus mirablis*, and *Bacillus subtilis*. These samples cannot rely on traditional culturing techniques and thus this integrated chip can be employed for rapid diagnostics of laboratory samples.

521c Determination of the Number of Bacteria Based on Autofluorescence on a Microfluidic Chip

Ning Bao[1], Balamurugan Jagadeesan[2], Arun K. Bhunia[2], and Chang Lu[3]. (1) Agricultural and Biological Engineering, Purdue University, West Lafayette, IN 47907, (2) Food Science, Purdue University, West Lafayette, IN 47906, (3) Agriculture and Biological Engineering, Chemical Engineering, Purdue University, West Lafayette, IN 47906

Bacterial counts are often determined for hygiene monitoring of products. In this study, we demonstrate the quantification of the number of bacterial cells based on the autofluorescence from the cell lysate on a microfluidic chip. We tested three model pathogenic bacteria (Listeria monocytogenes F4244, Salmonella Enteritidis PT1 and Escherichia coli O157:H7 EDL 933). In the experiment, a plug of ~150 pL containing lysate from 240 to 4100 cells was injected into a microfluidic channel with downstream laser-induced fluorescence detection under electrophoresis conditions. We found that the autofluorescence intensity increased with the number of cells almost linearly for all three bacteria. The autofluorescence remained a single peak when the cell lysate contained a mixture of different bacterial species. Our method offers a simple solution to the quantification of bacterial cells without any labeling, especially on a microchip platform. We believe that the method can be extended to other bacterial species.

521d A Microfludic Device For Flow Cytometric Analysis Of Innate Immunity Via TLR4 Signaling In Macrophage Cells

Nimisha Srivastava, James S. Brennan, Steven S. Branda, Anup K. Singh, and Amy E. Herr. Biosystems Research Department, Sandia National Laboratories, Livermore, CA 94551

Microfluidics provides unique technologies for exploring unprecedented realms in complex, yet fundamental, cellular analysis. One such realm that we are investigating is host pathogen interactions with single cell resolution. Macrophage cells, a key component of innate immunity, respond to microbial pathogens through membrane proteins called Toll-Like Receptors (TLRs). Specifically, TLR4 surface receptors identify lipopolysaccharide (LPS) present on the cell envelope of Gram-negative bacteria (e.g., Francisella tularensis, Yersinia pestis). Our presentation will report on the development of an integrated device that conducts complex cell preparation as well as flow cytometric analysis for real-time monitoring of key processes in TLR4 signaling.

Flow Cytometry is a high throughput tool for interrogating single cell – however, the different upstream cell handling steps involved in preparing cells for FC is a bottleneck. Sample addition, LPS dosage, mixing, multiple reagent addition for – temporal fixation of proteins, permeabilization of the cell membrane, labeling with phospho-specific antibodies, and selective washing steps are required. Our microfluidic platform integrates necessary cell handling steps with on chip fluorescence-based flow cytometry. The fused silica microfluidic device integrates mixers (t~15s), incubation chambers (~1000 cells per assay), surface modification for low cell adhesion and hydrodynamic focusing for Flow Cytometric detection (~10 cells/min). Multiple exchange of reagents is enabled through in situ photopolymerized porous acrylate filters (pore size ~4μm) that act as size-exclusion filters – allowing bulk pressure-driven fluid to flow through the monoliths, while retaining macrophages. Initial results from the assays are in agreement with conventional FC, with the compelling advantage of using only 10 μL of fluid and total analysis times of 15 min. We have also developed assays to detect the time-lapse phosphorylation of p38 (p38-P) and ERK (ERK-P), key proteins in TLR4 response. We will report detailed results from the streamlined flow cytometry assays and present details of our understanding of the innate immune response.

Web Page: roswell.ca.sandia.gov/srivastava.html

Microfluidic gDNA Quantification by Flow rate analysis

[1]*Rajesh Surapaneni*, [2]*Jungkyu Kim*, [1,2]*Bruce K. Gale*

[1] *Department of Mechanical Engineering, University of Utah, Salt lake City, UT*
[2] *Department of Bioengineering, University of Utah, Salt Lake City, UT*

Abstract

A system for DNA extraction and quantification was designed and fabricated using a nanoporous aluminum oxide membrane. In this system, as DNA was collected on the membrane, the flow resistance through the membrane increased, which was measured using a flow rate sensor. Based on this concept, an existing DNA extraction system was modified to quantify gDNA (genomic DNA) collected. The DNA extraction system was fabricated with PDMS micro-channels, double-sided tape and aluminum oxide membrane. The nanoporous membrane was embedded in a microfluidic system and DNA was collected over the membrane. To demonstrate the system, nanoparticles (370nm diameter) were used and extracted over the membrane of pore size 200nm. Over a period of time, particles were collected over the membrane with flow rates falling exponentially. After the proof of concept was tested, gDNA of 25ng/µl and 10ng/µl concentrations were extracted over 20nm pore sized membranes and similar results were obtained. The flow rate dropped drastically after 461.01±40.73 seconds of flow over the membrane for 25ng/µl and 1181.57±84.5 seconds for a 10ng/µl concentration using a flow rate sensor. For 10ng/ul and 25ng/µl DNA samples, the amount of DNA collected over the membrane for 461.01seconds and 1881.57seconds were 1312.857±93.99ng and 1280.6±113.13ng respectively. From this data it was evident that approximately 1300ng of DNA was required to make monolayer of DNA on nanoporous membrane. This experiment was repeated for various known concentrations of gDNA. An empirical graph was constructed with time on the x-axis and known concentrations on y-axis. With the empirical graph, gDNA sample concentration was possible to estimate from the time (value of first time derivative was maximum). By integrating with a microfluidic DNA sample preparation system, we can extract and quantify the DNA sample simultaneously.

Introduction

Various processes such as optical density (OD) measurements, fluorescent enhancement, ICP-OES[1] (inductively coupled plasma-optical emission spectroscopy, detects DNA based on stoichiometry of phosphorous within the molecule), and quantitative diagnostic assays[2,3] are generally used for quantifying DNA samples. These processes have drawbacks due to large initial costs or amount of time consumption and inaccuracies incurred due to fluorescent or UV absorbing impurities[4]. Labeling methods like fluorescent dyes provide alternate techniques in detection of dsDNA (double stranded DNA)[5]. In these methods extra cost is the need to buy fluorescence dye.

The purpose of this research was to design a microfluidic system for gDNA quantification by measuring the flow rates using a flow rate sensor (Nano Flow Sensor N-565, Upchurch Scientific®). DNA extraction system was fabricated with PDMS micro-channels, double-sided tape and aluminum oxide membrane[6,7,8]. To construct empirical graphs gDNA concentrations and time from the flow rate sensor when flow rate changes drastically due to

DNA collection over aluminum oxide membrane must be known. Known gDNA concentrations of 25ng/μl and 10ng/μl were chosen for experimentation. Flow rate measurement was done with some caution due to chances of inaccuracies introduced during measurement of time. The microfluidic DNA quantification method will be used for DNA quantity control with high throughput microfluidic DNA extraction system.

Methods and Material

The online-quantification system consists of a gDNA extraction chip with nanoporous membrane embedded in between two PDMS layers and a sensitive flow rate sensor. Preliminary experiments using nanoparticles were conducted using a 200nm pore size membrane whereas subsequent experimental results were obtained over 20nm pore size membrane due to size constraints. Extraction chips were manufactured by curing PDMS over sticky side of a 100μm thick double-sided tape (cover sheet not removed on the other side) in an oven at a temperature of 65°C. Holes of 2.25mm diameter were cored (using a coring tool) specifically to insert a 2.50mm OD tubing into the holes (tubing connects extraction system to a 3ml syringe (BD plastic Inc.) and flow rate sensor on either sides). Cover sheet over two such PDMS layers were removed and glued with the aluminum oxide membrane in between. This assembled structure was cured to strengthen the bond between two double sided tapes. This was done for about 3 hours in the same oven at a temperature of 65°C. The system was then ready to extract gDNA.

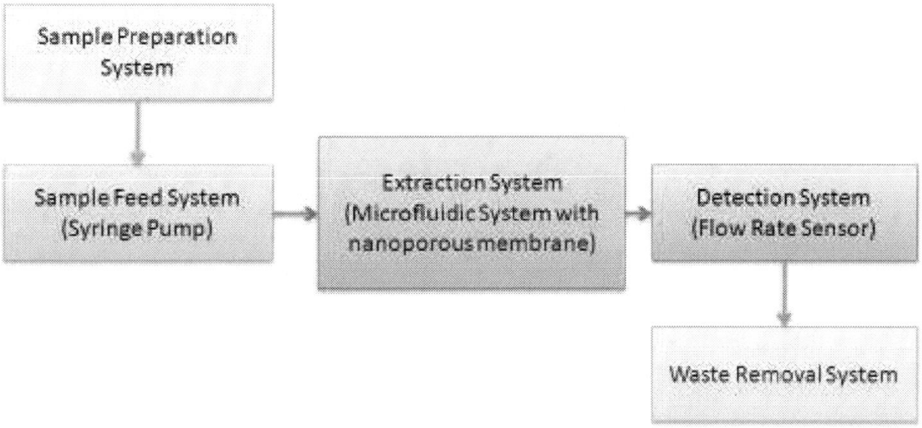

Figure 1: Schematics of experiment

gDNA samples were prepared by serial distribution of the concentration. The initial concentration of the gDNA sample was 50ng/μl. 0.5ml of this concentration was taken in a 1.5ml vile and then 0.2ml of the sample was separated into another vile and 0.3ml of TE buffer was added to the separated sample. The 0.3ml remaining sample in first vile was mixed with 0.3 ml of 1% of Triton X. This made the sample 50% dilute i.e., 25ng/μl concentration. The 0.5 ml of DNA sample in the second vile was again separated into 0.2ml and 0.3ml. 0.3ml was again mixed with 0.3ml of Triton X. The concentration of second sample was 10ng/μl. A 3ml syringe was mounted over a syringe pump, which injected gDNA to an extraction system at constant flow rate via connecting tube. The outlet of the PDMS Micro channel was connected to the flow rate sensor. As the DNA started collecting over the membrane, flow rate at the output of extraction system was reduced. This reduction in flow rate was constantly measured through a micro channel using a flow rate sensor. When the pore area of this membrane was fully blocked by DNA, flow rate dropped drastically. Flow rate data recording should be started simultaneously with the syringe pump. From these raw data, the data was modified and found the first derivative value and its time.

Results and Discussion

The developed online-quantification system was operated with nanoparticles and gDNA samples. Each concentration of gDNA sample was tested thrice with different extraction systems (of same membrane pore size). Data from the flow rate sensor was stored on a computer each time. This recorded data was then opened in excel. Data was tabulated in such a way that a single graph had data plotted with flow rates on y-axis and time on x-axis for different concentrations. Graphs were obtained as a result of 3 tests with each concentration as shown in figure 2. From figure 2, it was evident that flow rate drastically reduced as the membrane was completely blocked. A trendline was added to the end of the graph where the flow rate started to fall drastically. Trendline was used to determine the time at which the flow rate dropped drastically. Concentration and time were tabulated to plot empirical graph. The time taken for 0ng/μl concentration tends to infinity. Therefore concentration VS (1/time) was plotted as shown in figure 3. With more data points, a plot of concentration Vs time would yield more accurate curve, by means of which unknown concentration of gDNA can be interpolated on the graph. Quantification time for this system can be improved by reducing membrane area, thereby reducing the time for blockinging membrane. By reducing the area, sample volume for each experiment can be reduced as well as lesser concentration samples can also be extracted.

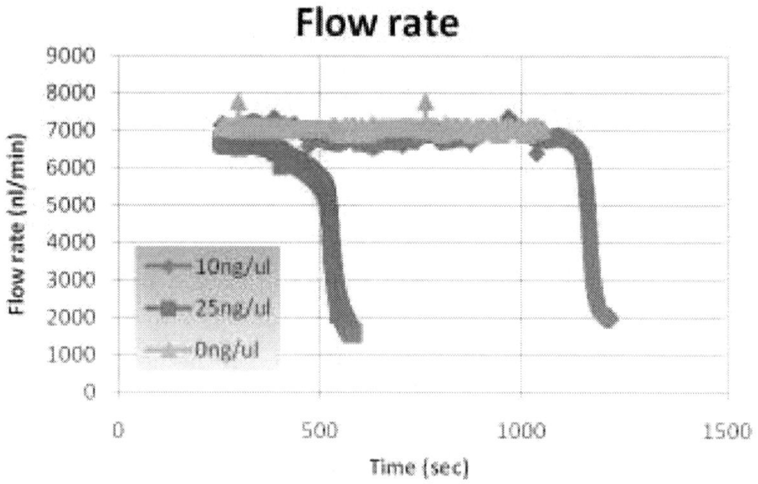

Figure 2: Graph for three different sets of experiments with three different concentrations (0ng/ul, 10ng/ul and 25ng/ul). The curve for a 0ng/ul is a straight line and there is no change in flow rate, whereas for other concentrations, flow rate drops drastically when the membrane is completely blocked with gDNA.

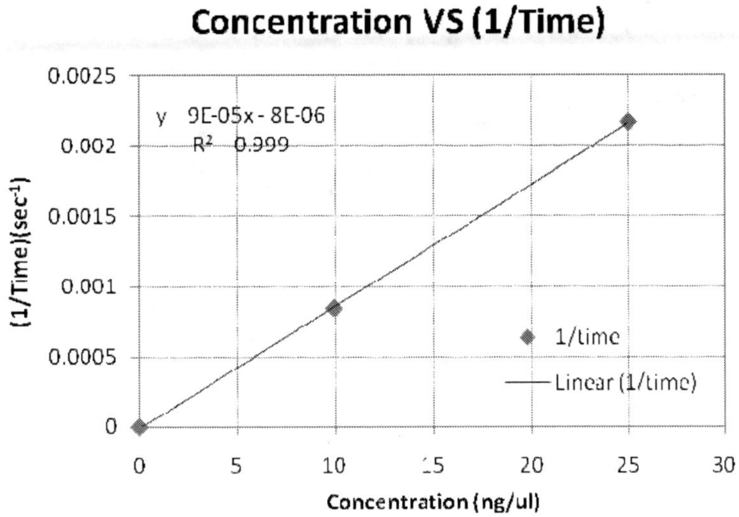

Figure 3: Empirical Graph with Concentration VS (1/Time). This Graph is used to know the concentration of an unknown gDNA sample by interpolating (1/time) where the flow rate drops drastically. A linear trendline with R^2=0.999, shows that the curve is a straight line.

To construct a cost effective system, a serpentine microfluidic channel (made of PDMS) is bonded at outlet of the extraction system. This serpentine microfluidic channel replaces a costly flow rate sensor. A dye can be mixed with the sample and using video analysis at the serpentine channel, flow rates can be determined. This assembly is much cheaper compared to the current system with a sensitive flow rate sensor.

Conclusion

Online quantification system developed in this study is based on the flow rate analysis. The present system includes extraction and quantification using aluminum oxide membrane and sensitive flow rate sensor. When integrated with a microfluidic DNA sample preparation system, this system may be useful to quantify gDNA from blood to good approximation. Knowing the concentration helps in knowing the amount of sample required during amplification. Also the system is cheap and easy to integrate. Future works include integration of this system with a microfluidic DNA sample preparation system and quantify gDNA in blood sample.

References

1. Claire A. English, Sheila Merson and Jacquie T. Keer, *Anal. Chem.* 2006, *78, 4630-4633.*
2. Gal, S.; Fidler, C.; Lo, Y. M.; Taylor, M.; Han, C.; Moore, J.; Harris, A. L.;Wainscoat, J. S. *Br. J. Cancer* 2004, *90*, 1211-15.
3. Zhong, X. Y.; Holzgreve, W.; Hahn, S. Hypertens. *Pregnancy* 2002, 21, 77-83.
4. Cavaluzzi, M. J.; Borer, P. N. *Nucleic Acids Res.* 2004, *32*, e13.
5. Singer, V. L.; Jones, L. J.; Yue, S. T.; Haugland, R. P. *Anal. Biochem.* 1997, *249*, 228-38.

6. JungKyu Kim, Karl Voelkerding, and Bruce K. Gale, "Patterning of an nanoporous membrane for multi-sample DNA extraction," *Journal of Micromechanics and Microengineering*, Vol. 16, pp. 33-39, 2006.
7. JungKyu Kim and Bruce K. Gale, Microfluidic DNA Extraction Using a Patterned Aluminum Oxide Membrane, in *Proc. Of SPIE: Microfluidics, BioMEMS, and Medical Microsystems IV*, San Jose, CA, January 23-28, 2006.
8. Jungkyu Kim, Bruce K. Gale, "Multi-DNA Extraction Chip Based on an Aluminum Oxide Membrane Integrated into a PDMS Microfluidic Structure," in *Proc. Of IEEE-MMB* 2005, *Oahu, Hawaii, May 12-15, 2005,* pp. 5-7.
9. Chih-Sheng Johnson Hou, Nebojsa Milovic, Michel Godin, Peter R. Russo, Raj Chakrabarti and Scott R. Manalis, *Anal. Chem.* 2006*, 78, 2526-2531.*
10. Lori Beach, Claude Schweitzer and J. C. Scaiano, *Org. Biomol. Chem.*, 2003***,*** *1, 450-451.*
11. Mark A. Burns, Brian N. Johnson, Sundaresh N. Brahmasandra, Kalyan Handique, James R. Webster, Madhavi Krishnan, Timothy S. Sammarco, Piu M. Man, Darren Jones, Dylan Heldsinger, Carlos H. Mastrangelo, David T. Burke. *Science 282, 484 (1998).*
12. M. P. Nandakumar, E. Nordberg Karlsson, B. Mattiasson, "Integrated Flow-Injection Processing for On-line Quantification of Plasmid DNA During Cultivation of E. coli." *BIOTECHNOLOGY AND BIOENGINEERING, VOL. 73, NO. 5, JUNE 5, 2001.*

521f Microfluidic Enzyme-Linked Immunosorbent Assay For Rapid Detection Of Salmonella Typhimurium In Surface Enhanced Poly(Methyl Methacrylate) Microchannels

Wei-Cho Huang[1], Yunling Bai[2], Shang-Tian Yang[3], and Ly James Lee[3]. (1) BioLOC LLC, 1381 Kinnear Road, Columbus, OH 43212, (2) Chemial and Biomolecular Engineering, The Ohio State University, Room 125A, 140 W. 19th Ave., Columbus, OH 43210, (3) Chemical and Biomolecular Engineering Department, The Ohio State University, 140 w. 19th Ave, Columbus, OH 43210

An enzyme-linked immunosorbent assay (ELISA) was performed in a microchannel (140 micron width by 125 micron depth) on a poly(methyl methacrylate) (PMMA) microchip. The increased positive charge and amine content on poly(ethyleneimine) (PEI) treated PMMA surface greatly enhanced antibody binding on the microchip, resulting in 3-fold enhancement in the fluorescence signal. With the reduced volume in the microchannel, less sample and reagent (<5 microliter) were consumed, and the incubation time for antigen and antibody binding for the ELISA was also significantly reduced to less than 1 h in each step. Total assay time was reduced to ~2.5 h, from 13 h for ELISA in conventional microtiter plates. The microfluidic ELISA can detect as few as 30 cells of Salmonella typhimurium (with an S/N ratio >7), which is 1-3 orders more sensitive than conventional ELISA methods performed on 96-well and 384-well plates. With the reduced reagent use, enhanced sensitivity, and easy and rapid operation, this microfluidic ELISA assay provides a low-cost alternative to conventional ELISA on 96-well and 384-well plates. Further automatic processing and parallel detection of multiple foodborne pathogens can also be achieved by adopting this platform in a multichannel or microarray format. This microchip can be readily integrated with a low-cost, portable detection system for various field applications, including food safety surveillance, environmental monitoring, and diagnostics for home healthcare.

521g Fabrication And Characterization Of A Soft-Polymer Piezoelectric-Actuated Peristaltic Micropump For Integration Into An Online Capillary Electrophoresis Instrument, And Use In Direct-Sampling Probe Neuroscience Experiments

Neil J. Graf and Michael T. Bowser. Chemistry, University of Minnesota, 207 Pleasant St. S.E., Minneapolis, MN 55455

With the growing demand for lab-on-a-chip devices each year, it is critical that resources be invested into the development of new and improved microstructures that are capable of moving and manipulating fluids in a controlled manner throughout these devices. Micropumps and microvalves are the structures that perform these operations, and are critical components to overall device performance. For this work, successful fabrication of a robust peristaltic micropump was performed. The performance of this micropump demonstrates it has great potential for integration into a variety of instrumental and microfluidic device platforms.

Fabrication of this micropump was performed using soft lithography techniques. A novel fabrication approach was used for construction of the microchannels. In brief, PDMS is cured over a raised structure, created by molding aluminum foil over a 75 μm diameter wire preadhered to a glass substrate. The resulting microchannel cross-sectional geometries have a rounded, Gaussian curve-like shape, and thus are ideal for complete valve shutoff. A membrane of thickness ~125 μm seals the channel from the outside world. Actuation of the micropump is achieved with piezoelectric cantilever motors. Adhered to the tip of these motors are precision machined aluminum edge masses that serve as valves for the micropump. When three signals with different phases are sent to the cantilevers, a wave-like motion creates a net flow of fluid through the device.

A substantial amount of characterization data for this micropump will be presented. The dynamic range for flow rates in this device is large, ranging from ~50-320 nL/minute. Different flow characteristics were observed depending on the phase of the input signal delivered to each of the respective cantilever motors. Net pumping of fluid occurs against backpressures recorded as high as 36,759 Pa, using an electronic pressure sensor. Valve performance data for this device will also be presented, including valve closing/opening characteristics and valve leakage. This micropump was developed for inline integration of a direct sampling probe with online capillary electrophoresis analysis, for the purpose of studying neurochemistry in rodent brains. Current efforts are underway to optimize this micropump before being fully integrated into the online capillary electrophoresis instrument.

569a Isoelectric Focusing In Contraction-Extraction Microchannels

Jaesool Shim[1], Prashanta Dutta[1], and **Cornelius F. Ivory**[2]. (1) School of Mechanical and Materials Engineering, Washington State University, Pullman, WA 99164-2920, (2) School of Chemical Engineering and Bioengineering, Washington State University, Pullman, WA 99164-2920

Microchip isoelectric focusing (IEF) is becomming more popular due to its short operation time, low cost, high resolution, low reagent consumption, ease of integration with various unit operations, etc. Until recently, microchip IEF had only been simulated in one-dimension which is not sufficient to model the behavior of proteins in complex microgeometries. In this study a two-dimensional numerical model is developed to simulate ampholyte-based IEF in both straight and contraction-expansion microchannels in the range, 6.21 The results of the simulation show that most of the ampholytes initially form pairs of peaks on the anodic and cathodic edges of the channel, which then move toward the channel center. A very steep pH gradient is initially formed at each electrode and gradually spreads to the middle of the channel as focusing proceeds. The numerical results reveal different focusing efficiencies and resolution in the narrow and wide sections of the contraction-expansion channel.

569b Multidimensional Isotachophoretic Focusing in Microfluidic Channels

John S. Paschkewitz, Joshua I. Molho, Hui Xu, Rajiv Bharadwaj, and C. Charles Park. Caliper Life Sciences, 605 Fairchild Drive, Mountain View, CA 94043

Isotachophoresis (ITP) is a powerful electrophoretic technique for sample concentration and separation; when used on microfluidic chips it may be necessary to use serpentine passages or expansion sections to meet design requirements. We have observed novel two-dimensional focusing effects in these geometries. Using a simple analytical model for the behavior in a turn, we illustrate that the focusing effect occurs because the spatial distortion of the leading buffer-trailing buffer interface generates a wall-normal electric field. Using this analytical model we identify the controlling non-dimensional parameters. We present the results of a computational parametric study of this focusing behavior in turns as well as a comparison to experimental visualizations. Finally, we present preliminary computational results for ITP in expansions.

569c Ionic Separation In Nanofluidic Channels

Xiangchun Xuan, Mechanical Engineering, Clemson University, 239 Fluor Daniel Engineering Innovation Building, Clemson, SC 29634-0921

This article presents a theoretical study of the efficiency, selectivity and resolution of ionic separation in nanofluidic channels. The separation of ionic and neutral solutes can be implemented in both pressure-driven and electric field-driven nanochannel flows. Such separation is attributed to the strong electrostatic interactions between ionic solutes and inherently charged channel walls.

614e DNA Dynamics When Electrophoresing in Microfluidic Post Arrays: Brownian Dynamics Simulation and Stochastic Modeling

Patrick S. Doyle, Aruna Mohan, and Ju Min Kim. Department of Chemical Engineering, MIT, Cambridge, MA 02139

The size-based separation of electrophoresing DNA chains of various lengths in microfabricated obstacle arrays (Volkmuth and Austin, Nature 1992) and self-assembled arrays of superparamagnetic beads (Doyle et al., Science 2002) has recently been experimentally established. The chain dynamics in these devices have been modeled previously (Minc et al., Phys. Rev. Lett. 2005; Dorfman, Phys. Rev. E 2006). However, previous studies do not fully take account of the electric-field dependence of chain extension that is observed in experiments (Randall and Doyle, Macromolecules 2006). In this study, we extend the continuous-time random walk model of chain dynamics in the obstacle array with account for incomplete chain extension. In addition, we evaluate the accuracy of the model by performing Brownian dynamics simulations of DNA chains of different lengths in a self-assembled array of magnetic beads at various electric field strengths. We provide predictions of chain mobility and dispersivity in the device, and the resolution achievable among chains of different lengths.

569e Multiple Regimes of Collisions of a Single Electrophoretically Translating Polymer Chain against a Thin Post

Sean P. Holleran and **Ronald G. Larson**. Chemical Engineering, University of Michigan, 3074 HH Dow / 2300 Hayward Street, Ann Arbor, MI 48109

We use a previously developed bead-spring Brownian dynamics model for simulating the topological interactions between polymers and thin obstacles to study electrophoretically translating DNA strands interacting with an immovable post over a wide range of chain lengths and field strengths. The field strength is quantified by the Peclet number based on the Kuhn length, which is the ratio of the rates of field-induced polymer motion to diffusive transport at the level of the Kuhn segment length. We quantify the effect of the post by using the "delay distance" between the position of the post after its encounter with the post and the position it would have had if it had translated at uniform rate in the absence of the post. We find that the "delay distance" induced by the entanglement interaction increases with higher fields, encompassing four distinct regimes. Two regimes exhibit the classic rope-and-pulley dynamics. One of these is a very high field strength regime in which the dimensionless delay distance induced by interactions with the post reaches its upper limit, and the other is a moderately high field strength regime in which the chain is not able to fully extend. In the two slower regimes, the polymer retains a coil-like shape as it diffuses laterally and eventually clears the post without deforming. We develop models that describe both the average delay and the distribution of delays for all regimes, except the slowest one, which is distinguished by a peculiar fractional power law relationship between delay distance and Peclet number. We also develop models that quantititively predict the distribution in decay distances and hence are of relevance to band broadening and separation efficiency.

614a Transport Of Current, Fluid And Analytes In Si/SiO_2 Fluidic Nanochannels

Yi Zhang[1], Alexander Neumann[2], and **Dimiter N. Petsev**[1]. (1) Department of Chemical & Nuclear Engineering, UNM, 209 Farris Engineering Center, Albuquerque, NM 87131, (2) Center for High Technology Materials, UNM, 1313 Goddard SE, Albuquerque, NM 87106

Fluidic nanochannels are often fabricated on Si/SiO_2 substrates, using methodology similar chip manufacturing. While such approach is very convenient and allows for a substantial reduction in the channel size, a potential problem could stem from the fact that the deposited SiO_2 layer becomes conductive when in contact with aqueous solutions. Hence if electroosmosis/electrophoresis is attempted in such channels, a significant part of the current could leak across the SiO_2 layer into the Si instead of being transported along the channel by the electrolyte solution. The effect of such current leakage is that the electric field in the fluidic channel becomes non uniform, which will inevitably affect also the transport of fluid and analytes.

We present an analysis of the system Si/SiO_2/Electrolyte solution based on numerical solution of transient Maxwell electrodynamics equations. The fluidic channel with Si/SiO_2 walls is the considered as a leaky capacitor (see the Figure). Our data show that there is a relationship between the SiO2 layer thickness and the channel length above which the current leaks become significant. Generally longer channels lead to greater current leaks.

While a potential problem, the current leaks could be utilized to modify the field shape in the fluidic nanochannel. For example it is possible to create local minima or maxima in the potential distribution along the channel which could be utilized for controlled analyte transport or focusing and separation.

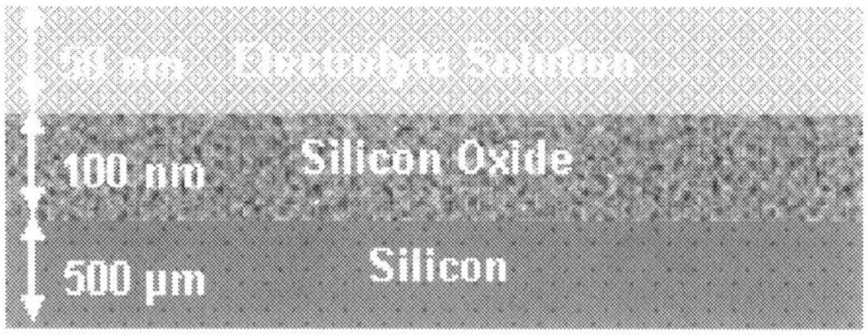

614b Field Effect Flow Control In Pdms Microchips

Prashanta Dutta and Keisuke Horiuchi. Mechanical Engineering, Washington State University, Pullman, WA 99163

In Lab-on-a-Chip microfluidic devices, electroosmotic flows (EOF) are most widely used to drive samples. The magnitude of EOF velocity in a microchip can be controlled either by changing the external electric field applied along the flow direction or by modifying the electrokinetic potential at the Stern plane (also known as zeta potential). It is relatively easy to control the overall flow rate in a straight microchannel by adjusting the external electric field applied in the flow direction. This kind of control is generally global, and it affects the entire electric field region. However, it cannot be applied to a local area of interest or in complex micro-geometries. But, the selective manipulation of zeta potential has promises to control the flow locally. To test this hypothesis, a field-effect transistor is developed to alter the zeta potential in microfluidic chips. Here, microchannel structures and field-effect transistors are formed on poly-di-methyl siloxane (PDMS) using soft lithography techniques, and a micro particle image velocimetry technique is used to obtain high resolution velocity distribution in the controlled region. The flow control is observed at relatively low gate voltage (less than 50 [V]), and this local flow control is primarily due to current leakage through the interface between PDMS and glass layers. A leakage capacitance model is introduced to estimate the modified zeta potential for the straight channel case, and an excellent agreement is obtained between the predicted and experimental zeta potential results. This leakage-current based field-effect is then applied to a T-channel junction to control flow in the branch channel. Experiments show that the amount of discharge in the branch channel can be controlled by modulating gate voltage.

614c Electokinetic Energy Conversion In Fluidic Nanochannels

Chris Davidson and Xiangchun Xuan. Mechanical Engineering, Clemson University, 102 Fluor Daniel Engineering Building, Clemson, SC 29634-0921

This article presents a theoretical analysis of the efficiency and output power of electrokinetic energy conversion in fluidic nanochannels. Phenomenological equations are utilized to clarify thermodynamically the conversion principle between mechanical energy and electrical energy in electrokinetic flow. Such electrokinetic devices performance is gauged by a dimensionless "figure of merit" Z, a combination of phenomenological coefficients. We demonstrate that Z is dependent on four non-dimensional parameters: electrokinetic height, zeta potential, Levine number and Dukhin number.

614d Electrokinetic Instabilities In Non-Dilute Colloidal Suspensions

Guru Navaneetham, Mechanical and Aerospace Engineering Department, Micro/Nanofluidics Laboratory, Arizona State University, Tempe, 1711, South Rural road, ECG346, Tempe, AZ 85281 and Jonathan D. Posner, Mechanical and Aerospace Engineering, Arizona State University, Tempe, AZ 85287-6106.

ABSTRACT

The present work focuses on the experimental study of electrokinetic instabilities in non-dilute colloidal suspensions of nanometer scale polystyrene spheres. The addition of charged particles to a solution increases the conductivity of the solution due to the diffuse electrical double layer of ions surrounding each particle. Electrokinetic flows with conductivity gradients become unstable when the electroviscous stretching and folding of conductivity interfaces grow faster than the dissipative effects of molecular diffusion. In this work, a conductivity gradient is achieved at the intersection of a Y-shaped PDMS microchannel using pressure driven flows. Experimental scalar images, critical electric fields required for instability, and mean scalar energy perturbation fields over a wide range of applied D.C electric fields and conductivity ratios will be presented. Variation in the critical electric field and scalar perturbation values as a function of the pressure driven flow velocity will also be presented. These instabilities may be important in electrokinetic manipulation of non-dilute colloidal suspensions such as human blood, cells, material synthesis, or pharmaceutical development.

569d DNA Stretch during Electrophoresis Due to a Step Change in Mobility

Patrick T. Underhill, Chemical Engineering, University of Wisconsin-Madison, Madison, WI 53706 and Patrick S. Doyle, MIT, Room 66-456, Dept. of Chemical Engineering, 77 Massachusetts Ave, Cambridge, MA 02139.

Efficient stretching of DNA in microfluidic devices is important for new genomic mapping methods such as with Direct Linear Analysis. Two aspects of these devices that are common are the transient nature of the stretching and a change of the strain rate over the scale of the molecule. We have investigated the stretching of DNA undergoing electrophoresis across a step change in mobility, which is a simplified system which has these two properties. This system represents a limiting regime of experiments and allows us to use a toy model to identify the key dimensionless groups, an effective Weissenberg number and Deborah number. Using the toy model we determined the scaling of the stretch with those groups and verified the results using Brownian dynamics simulations. Key features of the stretching are that there is no coil-stretch transition and the finite extensibility only affects the response if the chain is stretched to a significant fraction of the contour length.

652a Capillary Electrophoresis Single-Strand Conformation Polymorphism (Ce-Sscp) For Active Community Profiling Of Complex Microbial Communities

Sage R. Hiibel[1], Amy Pruden[2], and Kenneth F. Reardon[1]. (1) Chemical and Biological Engineering, Colorado State University, 100 Glover Building, 1370 Campus Delivery, Fort Collins, CO 80523-1370, (2) Civil and Environmental Engineering, Colorado State University, 1320 Campus Delivery, Fort Collins, CO 80523-1370

Current community profiling techniques, such as T-RFLP, DGGE, and TGGE, fail to distinguish active from inactive microbial species in a mixed community. The purpose of this study was to validate the use of active community profiling (ACP) for characterizing the active microbial populations of constructed laboratory mixed cultures and a field-scale bioreactor treating acid mine drainage (AMD). ACP involves monitoring the composition and activity of a mixed microbial culture via comparative measurements of 16S rRNA and rDNA using capillary electrophoresis single-strand conformational polymorphism (CE-SSCP). The cellular ratio of total RNA to DNA is known to be proportional to growth rate in a variety of organisms. CE-SSCP involves denaturing of double-stranded PCR products that are rapidly cooled to form unique single-stranded conformations based on their nucleic acid sequences. Capillary electrophoresis is then used to separate the conformations based on electrophoretic mobility, resulting in an electropherogram with a series of peaks representing dominant members of the microbial community. Species identification is then determined by comparing the electrophoretic mobilities of individuals from a clone library to the individual peaks of the community profile.

The first stage of the study demonstrated that the rRNA:rDNA ratio obtained from CE-SSCP analysis can differentiate between different growth phases of a model organism, *E. coli* K12. Additionally, the ability to differentiate between different species was established. For the pure culture studies, *E. coli* was harvested from a chemostat at four different growth rates (μ•. The rRNA:rDNA for each μ value was calculated using three independent methods: (1) CE-SSCP, (2) quantitative DNA marker using traditional slab gel electrophoresis, and (3) spectrophotometrically. For each analytical method, increased growth rate was found to correlate with a larger rRNA:rDNA.

Pure cultures of four species, *Desulfovibrio salexigens*, *Desulfotomaculum ruminis*, *E. coli* K12, and *Pseudomonas putida* F1, were combined to create simple mixed cultures. The mixed communities were exposed to perturbations known to inhibit the growth of selected species. The perturbations included oxygen content, salt concentration, and the addition of chlorhexidine. For each disturbance to the community, a quantifiable decrease in the rRNA:rDNA of the species targeted for inhibition was detected using CE-SSCP, while the rRNA:rDNA of the non-targeted species remained relatively constant. These studies validated the use of CE-SSCP as an appropriate means of characterizing the active members of a mixed microbial culture.

When applied to a field-scale bioreactor treating AMD, a more complex and diverse microbial community was observed in the rDNA profiles than in the rRNA profiles, indicating that many of the species present in the system are not active. Species representative of the three main groups of microorganisms associated with sulfate-reducing bioreactors (cellulose degraders, fermenters, and sulfate reducers) were more dominant in the rRNA profiles, verifying their active role in the remediation process. In addition to the three main groups, several species with unknown function were also prevalent in the rRNA profiles. These species may be actively involved in AMD remediation, or may be particularly suited to the environment of the bioreactor.

The results of this study validate the use of ACP as a tool to characterizing the active members of mixed microbial cultures. The method gives additional information regarding a community beyond that obtained from standard 16S rDNA community profiling techniques. The differences observed between standard 16S rDNA profiling and the newly developed rRNA-based ACP indicate that conclusions based on DNA measurements alone may not accurately depict the microbial community. By identifying the active species, ACP is an important step forward in fully understanding complex microbial communities.

652b Advances In DNA Sequencing And Forensic Sizing By Free-Solution Conjugate Electrophoresis In Microfluidic Devices

Jennifer A. Coyne[1], Robert J. Meagher[2], Jennifer S. Lin[1], Russell D. Haynes[3], and Annelise E. Barron[1]. (1) Chemical and Biological Engineering, Northwestern University, 2145 Sheridan Road, Evanston, IL 60208, (2) Sandia National Laboratories, P.O. Box 969, MS 9292, Livermore, CA 94551, (3) Chemistry, Northwestern University, 2145 Sheridan Road, Evanston, IL 60208

Recent advances have been made in the development of Free-Solution Conjugate Electrophoresis (FSCE) as a method for long-read DNA sequencing and rapid forensic sizing ("DNA fingerprinting") in microfluidic devices. Size-based separations of DNA molecules are critically important to current methods of DNA sequencing, genotyping and forensic analyses; these separations are presently achieved by the use of a polymeric sieving matrix that sorts the DNA fragments in order from smallest to largest. Achieving size-based separations of DNA molecules on microchips in free-solution electrophoresis with no polymer matrix will eliminate the time-consuming, difficult step of loading viscous polymeric matrices into the glass or plastic devices. Additionally, there are many speed and cost advantages to performing electrophoretic separations in miniaturized "lab-on-a-chip" devices, which promise higher throughput and potentially, a lower per-run cost for genetic analysis.

In FSCE, DNA primers are conjugated end-on to a monodisperse polyamide "drag-tag" that subsequently enables the size-separation of DNA in free solution. The drag-tags effectively act as hydrodynamic "parachutes," breaking the linear scaling of the charge-to-friction ratio of DNA in electrophoresis. If each DNA molecule is modified by the addition of a single, monodisperse polyamide drag-tag, one unique peak results for each uniquely sized DNA molecule, and high-resolution separations of a mixture of different-sized DNA molecules can be accomplished by microchannel electrophoresis without a polymeric gel or sieving matrix. Both non-natural, highly repetitive, genetically engineered polypeptide drag-tags and poly-*N*-substituted glycine ("peptoid") drag-tags synthesized in our lab have been used to separate DNA fragments for sequencing, genotyping and forensic analysis applications.

High-resolution FSCE separations have been previously hindered by the lack of suitable, monodisperse drag-tags. Newly developed polypeptide drag-tags, however, have been used to separate DNA sequencing fragments with substantially higher resolution and cleaner results than previously reported for this method. The full potential of rapid FSCE separations of DNA sequencing fragments is demonstrated on microfluidic chips, where sequencing fragments are resolved in approximately 110 seconds. In addition to sequencing, drag-tags have also been used to separate DNA fragments for genotyping and "DNA fingerprinting" forensic applications. A highly multiplexed single-base extension (SBE) assay using peptoid drag-tags was developed to genoptype point mutations in exons 5-9 of the p53 gene, a gene associated with many types of human cancer; sixteen point mutations are analyzed in under 70 seconds by microchip electrophoresis in free solution. Preliminary "DNA fingerprinting" analysis by FSCE using the "miniSTR" primers developed by Butler *et al.* (*J. Forensic Sci.* 2003) has also been developed to size fragments up to 150 bases in length. Several challenges essential to further successful, rapid FSCE analyses on microchips are also currently under investigation, including production of significantly larger drag-tags and development of a novel, post-PCR bioconjugation method. Overall, the transition of FSCE to microfluidic devices is greatly simplified by the elimination of the need for viscous sieving matrices, and this technology promises to be easily incorporated into fully integrated microfluidic bioanalysis devices.

652c DNA Nanofiltration Far From Equilibrium

Nabil Laachi[1], Carmelo Declet[2], Christina Matson[3], and Kevin D. Dorfman[1]. (1) Chemical Engineering and Materials Science, University of Minnesota, 151 Amundson Hall, 421 Washington Ave SE, Minneapolis, MN 55455, (2) University of Puerto Rico, Mayaguez Campus, Bosque st. #58 Apt. 305, Mayaguez, PR 00861, (3) Biological Engineering, Mississippi State University, MS 39762

Nanofilters are a promising new class of nanofabricated devices that have proven very successful at rapidly separating sub-kilobasepair DNA and denatured proteins using weak electric fields, typically less than 100 V/cm. The device consists of a series of narrow slits (~50nm) connecting deep wells (~300nm) that trap the DNA. The rate of escape from the wells is a function of molecular weight, leading to a size-dependent elution process. Experiments and near-equilibrium theoretical arguments both indicate that smaller DNA emerge from the device first, and that separations fail at ~100V/cm. We show, on theoretical grounds, that the range of useful electric fields in nanofilters should extended beyond the experimentally observed limit of 100V/cm to high values of around 1000 V/cm, which would enhance the speed and the performance of this class of DNA separation devices. Based on scaling arguments, the separation of short, rod-like DNA molecules at these high fields occurs by a new mechanism that we call "torque-assisted escape". Although DNA migrate from slit to slit in a strong field without much translational diffusion into the well, they can arrive at the slit entrance in a forbidden orientation due to rotational diffusion. Their escape is then enabled by the torque that originates from the non-uniform electric field at the slit entrance. The quadratic dependence of the torque on the molecule size indicates that larger molecules will now emerge first. As a result, we predict that separations in this high-field regime operate in band-inverted manner; larger molecules should now elute first. This claim is further examined via Brownian dynamics simulations. The simulation results confirm that the mobility increases with molecular weight, and also indicate a quasi-plateau in the mobility at very large fields. In an effort to optimize the nanofilter geometry for separating short DNA, we also study the effect of the well depth on the elution process. As the well deepens, short molecules will span smaller gradients in the electric field, which can delay their exit by torque-assisted escape. The size-dependent optimal well depth should then lead to a method for a better tuning of the separation

652d Designing Polymer Systems To Minimize Peak Widths During Electrophoretic DNA Separations In Microfluidic Devices

Christopher P. Fredlake, Daniel G. Hert, and Annelise E. Barron. Department of Chemical and Biological Engineering, Northwestern University, 2145 Sheridan Road, Evanston, IL 60208

Electrophoresis of DNA molecules through entangled polymer solutions in both capillaries and microchips continues to be an invaluable tool for genetic analysis and is used in many genotyping and forensics applications as well as DNA and genomic sequencing. The transfer of most of these separation methods fully to microchips has already been accomplished or is in the process of being implemented. The design of the cross injector in microchips has greatly increased the efficiency of these separations, therefore reducing the required separation distance for many of these assays. To fully optimize the resolution of these separations, the polymer matrix and wall coating must be designed in such a way that both improves the peak spacing while also reducing the widths of the peaks. Separation mechanism models have been developed to describe the size-based mobility of DNA in the matrices. However, these models, as well as most experimental studies, generally ignore band broadening effects, which actually limit resolution for most separations. The narrow injection zones in microchips improve upon the broad injection plugs found in capillary-based separations. While injection plug widths dominate peak widths during capillary separations, microchips tend to be limited by analyte-wall interactions and electric field induced dispersion. Interestingly, different polymer matrix chemistries give rise to vastly different band broadening properties in the system. For example, we have shown that peaks are generally much wider in linear polyacrylamide based matrices (which are used ubiquitously in the literature) than in poly(N,N-dimethylacrylamide) matrices. In this report, we discuss both the physical and chemical properties of the polymer matrix along with the electrophoresis conditions that affect the widths of the analyte peaks during separation. Specifically, we show the quantitative effect of polymer and matrix physical properties on the DNA dispersion coefficients. Additionally, we explore the effect of wall coating chemistry on the peak widths as well and discuss the design of a DNA separation system to minimize band widths and maximize resolution for DNA separations.

652e Dielectrophoretic Detection of of Hybridized Genetic Bead Suspensions Using a Novel Suspension Kaleidoscope Approach

Zachary R. Gagnon[1], Jason Gordon, and Hsueh-Chia Chang[2]. (1) Chemical and Biomolecular Engineering, University of Notre Dame, 182 Fitzpatrick Hall, Notre Dame, IN 46556, (2) Department of Chemical and Biomolecular Engineering, University of Notre Dame, 181 Fitzpatrick Hall, Notre Dame, IN 46556

It is well known that silica particles on the order of 100 nm have an electrical conductivity < 10^{-7} S/m, much less than even that of deionized water, ~ 20 μS/m, still exhibit positive dielectrophoresis (DEP) when suspended in such a low conductive electrolyte. Such behavior is commonly attributed to Stern layer adsorption of surface charges. This increased surface charge density on the particle surface facilitates electron transport around the particle and can thus give rise to an increase in the effective conductivity of the particle enabling low conductive particles to still exhibit positive DEP in a media of greater conductivity.

The frequency at which the induced particle dipole goes to zero, known as the crossover frequency, is highly dependent on the surface conductance of the particle. In this work we consider the effect that DNA hybridization has on the surface conductance of a silica particle of varying diameters ranging from 100 nm – 1 im and exploit such detectable changes in order to rapidly detect DNA hybridization reactions. Instead of detecting particle crossover frequency through the tracking of individual particles, we take a novel DEP suspension pattern approach.

By designing a quadrupole micro-electrode with specific symmetries and length scales comparable to the average suspension screening length, we are able to overlap the different suspension aggregates into a kaleidoscope of two dimensional suspension patterns with different symmetries and anti-symmetries (See Fig. A-B). An open orifice exists at the center of the quadrupole for positive DEP and a closed circular aggregate appears for negative DEP. Unlike previous DEP colloidal results, the reported suspension pattern orifice dimensions are highly dependent on the field frequency and particle surface charge, and unlike traditional crossover frequency measurements, can be used to directly calculate the field induced DEP particle force.

Additionally, the detection of particle surface conductance typically requires the determination of multiple crossover frequencies of either particles suspended in various electrolyte conductivities, or of particles of varying sizes. It will be shown that the novel suspension pattern method allows accurate determination of surface conductance through the analysis of only one suspension pattern consisting of silica particles of varying size. Therefore, this novel suspension pattern approach allows one to determine the DEP force, the particle crossover frequency, and the particle surface conductance, and hence whether or not a particle surface has undergone DNA hybridization, from only one sample and in only a few seconds.

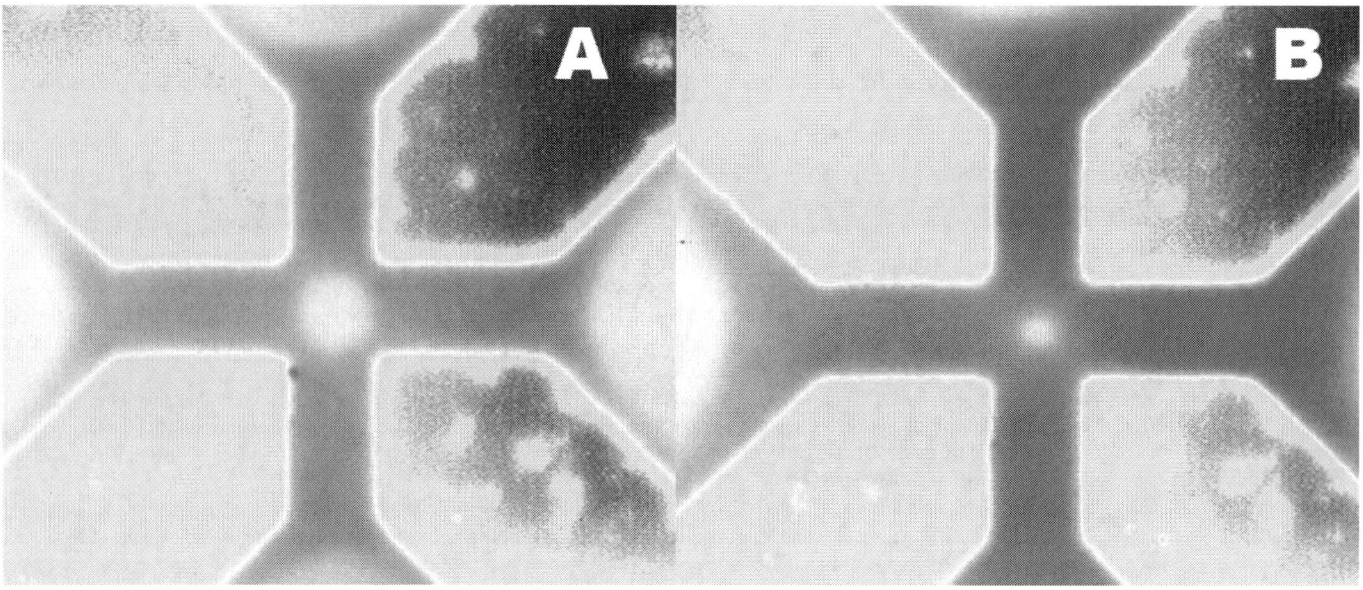
A
B

652f Dispersive Mixing Effect Caused By Combined Effect Of Channel Morphology And Electrophoretic Mobility In Poiseuille Flows

Rodrigo Hidalgo, Chemical Engineering, University Catolica del Norte, Avda Angamos 0610, Antofagasta, Chile, **Mario A. Oyanader**, Universidad Catolica del Norte, Avenida Angamos, Antofagasta, 0610, Chile, and Pedro E. Arce, Chemical Engineering, Tennessee Technological University, Department of Chemical Engineering, P.O. Box 5013, Cookeville, TN 38505.

Several electrokinetic processes (i.e. electro-assisted drug delivery, micro-electrophoretic separations, soil remediation, and material processing) can be optimized if the key parameters, i.e. driving factors for separation and mixing, are identified and manipulated to enhance the operation efficiency. The morphology of the capillary channel can modify the hydrodynamics of the system and the electrostatic characteristics of the walls can influence the electroosmosis. Collectively, these factors have an effect on the dispersive mixing phenomena. In this contribution, a fully developed Poiseuille flow has been subjected of study as to determine the effect of channel morphology combined with the electrostatic wall characteristics may have ultimately on the effective transport parameters of the system.

After the analysis of the electrohydrodynamics of the system has been determined, the molar species continuity equation is used to analyze the solute dispersion in the channel. Asymmetrical boundary conditions have been imposed over the system to capture the most representative behavior of the fluid. Several numerical examples illustrate the hydrodynamic and dispersion trends of the system using channel axial deformation, Debye Length, Electrostatic Potential ratio as possible design factors, for example.

652g Simulation Of DNA Hybridization Kinetics: From Single Target To Competitive Assay

Hong Susan Zhou, Chemical Engineering, Worcester Polytechnic Institute, 100 Institute Road, Worcester, MA 01609

We use a finite element software (Comsol Multiphysics) to study the kinetics of DNA hybridization and the potential limitation of the reaction rate caused by diffusion effects. The value of the parameters influencing the kinetics can sometimes span several orders of magnitude. Therefore, a wide range of values was used in order to cover the conditions seen in DNA microarrays. The competition between specific and non specific hybridization is also studied. In the cases where hybridization is diffusion limited, a convective transport is added to successfully overcome this limitation.

AUTHOR INDEX

AUTHOR INDEX

AUTHOR INDEX

AUTHOR INDEX

American Institute of Chemical Engineers
3 Park Avenue
New York, New York 10016-5991

ISBN: 978-1-60423-827-3